Leading Issues in Innovation Research

Edited by

Daniele Chauvel

Leading Issues in Innovation Research
Volume One
Copyright© 2011 The authors

First published September 2011 by
Second Print January 2012
Academic Publishing International Ltd, Reading, UK
http://www.academic-publishing.org

info@academic-publishing.org

ISBN: 978-1-908272-24-9

Note to readers.
Some papers have been written by authors who use the American form of spelling and some use the British. These two different approaches have been left unchanged.

Printed by Goodnewsdigitalbooks in the UK.

Contents

About the Editor

Danièle Chauvel is a Research professor at SKEMA, (Sophia Antipolis, France), and a member of the SKEMA Research Centre in Knowledge, Technology and Organization where she teaches Knowledge Management and Innovation. Her main expertise focuses on the human aspects of management and organizational dynamics: knowledge management, change management, team management, innovation management and leadership.

She has 12 years experience in academic research in Knowledge Management. She has developed theoretical and applied research in KM and built up several international research partnerships with academic and corporate institutions. She designed and taught several clusters on Knowledge Management and has regularly lectured on the MBA, Masters and Engineering programs in several French Business schools and Engineering schools.

She also works as a consultant in innovation and change management for the public sector including experience with a Finnish group. She previously had managerial responsibilities in pedagogical engineering and senior executive education in an international environment. She is active as a specialist in Knowledge Management in French think tanks and has acted as an expert in Knowledge Management on several EU projects.

Her research is dedicated to the evolution of Management principles, focusing on the role of Knowledge Management and its main trends and current development. She has authored and co-authored close to 50 articles, chapters and proceedings on Knowledge Management and co-edited the book, *"Knowledge Horizons: the Present and the Promise of Knowledge Management"*, with Charles Despres (Butterworth-Heinemann).

She holds a Master of Sciences in Social Sciences Research Methodologies from The Open University, UK and a MA in foreign languages.

List of Contributing Authors

Maria Antikainen, *Tampere University of Technology (TUT), Finland*

Liselore Berghman, *VU University Amsterdam, The Netherlands*

Juan-Gabriel Cegarra-Navarro, *Universidad Politécnica de Cartagena, Spain*

Gabriel Cepeda-Carrion, *Universidad de Sevilla, Spain*

Greg Clydesdale, *Massey University, Auckland, New Zealand*

Sinead Devane, *Bournemouth University, UK*

John Howard, *University of Central Lancashire, Preston, UK*

Daniel Jimenez-Jimenez, *Universidad de Murcia, Spain*

Alexandros Kakouris, *National and Kapodistrian University of Athens, Greece*

Albino Lopes, *ISCTE - Lisbon University Institute, Portugal*

Lawrence J. Loughnane, *CETYS Universidad Mexicali, Mexico*

Florinda Matos, *ISCTE - Lisbon University Institute and ESTG - Polytechnic Institute of Leiria, Portugal*

Nuno Matos, *PMEConsult, Portugal*

John D. Politis, *Higher Colleges of Technology, Dubai, UEA.*

Susana Rodrigues, *ESTG - Polytechnic Institute of Leiria, Portugal*

Eric Shiu, *The University of Birmingham, UK*

Kanji Tanimoto, *Hitotsubashi University, Tokyo, Japan*

Julian Wilson, *James Wilson (Engravers) Ltd, Poole, UK*

Introduction to Leading Issues in Innovation

Today *Innovation* is a focal point of attention because it is perceived as the organizational mainspring for gaining competitive advantage, enhancing performance and building successful futures. Innovation is not a new phenomenon; it dates from the ascent of humankind and can be understood as a way to organize human activity around the development of something new with economic value added.

Specifically defining the concept of innovation is not so simple, however. The earlier and even more recent literature displays numerous attempts to draw the conceptual contours of what *Innovation* means. The different perspectives employed by scholars and management gurus have led to a multifaceted concept. Innovations, according to Schumpeter (1939), consist of the practical implementation of knowledge, ideas, or discoveries and rely on entrepreneurial capabilities, not on inventiveness. Later (1942) he underscored the significant role of organization with its structural dimensions in the development of innovations. Peter Drucker states that, "Innovation is the specific tool of entrepreneurs, the means by which they exploit change as an opportunity for a different business or a different service" (1985, p.55). Tushman and Nadler (1996) insist more on the fundamental dimensions of the innovation itself, creation and diffusion / commercialization. Others, like Margaret J. Wheatley (1992) highlight the fostering role of, "...information gathered from new connections; from insights gained by journeys into other disciplines or places; from active, collegial networks and fluid, open boundaries"(1992, p.113). This last approach underlines the power of cognitive diversity in the creative process and presages the new forms of innovation developed in the Knowledge era.

Research in several academic fields has addressed the various aspects of innovation, including marketing, quality management, operations management, technology management, organizational behaviour, product development, strategic management and economics, with incomplete linkages in evidence across those fields (Hauser, 2006).

According to Wolfe (1994), "...several adequate, circumscribed, theories of innovation exist, but each applies under different conditions." This argues that research efforts should be directed at determining the contingencies between these various theories (Abrahamson, 1991; Eveland, 1991; Mohr, 1987; Poole and Van de Ven, 1989; Tomatzky and Fleischer, 1990; Van De Ven and Rogers, 1988). Wolfe suggests that three main streams of research would help organize the literature on the topic: diffusion of innovation, organizational innovativeness, and process theory ... each of which has been explored from different perspectives and dimensions.

The emergence of the Knowledge Economy, stemming these two past decades from a technological and socio-organizational break, puts special emphasis on knowledge assets, fast growing ICT developments, and the consequent digitalization of design and production. The Knowledge society recognizes knowledge as the main source of value, and knowledge creation and commercialization as the drivers of economic growth. In simple terms, from knowledge comes innovation which leads to the creation of new knowledge and consequently value creation. Knowledge and innovation are intertwined with knowledge being the catalyst for innovation.

From this perspective the concept of Innovation has broadened its horizon by transcending the boundaries of techno-scientific R&D, to embrace new *socio*-technical dimensions that include human, social and organizational factors. Historically, innovation was equated with the development of new products and technologies. In the Knowledge era, it is also applied to new services, business concepts, organizational processes, strategies, and management practices: "There is now a greater recognition that novel ideas can transform any part of the value chain" (Birkinshaw et al, 2011, p.43).

Moreover, the concept of innovation has spanned R&D laboratories and organization boundaries by extending to the whole organization and its environment, integrating the duality of internal vs external sources of innovation. This shift of focus has a direct impact on actors. In the knowledge society, any knowledge worker becomes a potential innovator: "...innovation has come to be seen as the responsibility of the entire organization." (Birkinshaw et al, 2011, p.43).

All these important changes entail critical issues for corporate management. Companies need to improve internal efficiency, bolster innovation and create innovation strategies in the face of an increasingly competitive marketplace. They must build up, on an enterprise-level, innovation infrastructures, practices and processes that deal with creativity, innovativeness, and innovative capabilities. They must also create and manage a context appropriate to innovation, including the requisite behaviour, knowledge and learning, collaboration and leadership factors (Van de Ven & Engelman, 2004; Desouza, et. al., 2009).

From a macro / environmental perspective another shift has emerged that affects the way in which innovation is conceived. The traditional model of "closed innovation," which implies internal resources and control, has become relatively unsustainable in the new economic reality. Now, in order to leverage innovation capacity, companies must find a way to tap into and harness new knowledge and ideas that reside beyond their boundaries (Chesbrough, 2002). "Open innovation" and related phenomena such as crowd sourcing have become increasingly prevalent.

Recently, new developments in collaboration technologies have contributed to the emergence of Web 2.0 applications which considerably increase connective and collaborative potentials. This leads naturally to Enterprise 2.0 (MacAfee, 2006) and Innovation 2.0 – *tapping un-tapped innovation talents* (Robson, 2007).

All these viewpoints illustrate that behind the term "*innovation*" exist an assembly of broad, complex and even chaotic processes that focus on renewal. This variety of approaches has showed that innovation is ascribed different characteristics - nature (incremental vs radical); object (product, service, process or business concept), level (intra/inter-organization, open innovation), which open onto many avenues within the field of innovation.

The spirit of this collection is to represent well the current developments in this field and its evolving multidimensionality in academic research and business practice. The selection favours works which are founded on solid theoretical footings and investigate new trends and/or specific niches that provide new perspectives in addressing the growing complexity and sometimes chaotic nature of the concept of Innovation. As noted above, the literature is growing, the concept expanding and this volume seeks to explore movements on the frontier as well as at the centre of Innovation as a field of study and practice.

The following chapters are derived from research articles published in the *Electronic Journal of Knowledge Management* and refereed proceedings from the *European Conference on Innovation and Entrepreneurship*. Taken together the chapters illustrate and deepen the different streams of thought noted above.

Daniele Chauvel
Skema Business School, Sophia Antipolis, France
September 2011

References

Abernathy, William J., Clark, Kim B, 1985. Innovation: Mapping the Winds of Creative Destruction, Research Policy. Amsterdam: Feb 1985. Vol. 14, Iss. 1; pg. 3.

Birkinshaw, Julian; Bouquet, Cyril & Barsoux, J.-L. 2011. The 5 Myths of Innovation. MIT Sloan Management Review; Winter 2011. pp 43-50.

Chesbrough, Henri, 2003. The Era of Open Innovation, MIT Sloan Management Review, Spring 2003, p 35-42.

Desouza,K.C.,Dombrowski,C.,Awazu,Y.,Baloh,P.,Sangareddy,S.R.P.,Jha,S.,Kim,J.Y. 2009. Crafting Organizational Innovation Processes. Innovation: Management, Policy &Practice, 11(1), 6–33.

Drucker, Peter, 1985, Innovation and entrepreneurship: practice and principles. HARPER & ROW, PUBLISHERS, New York Cambridge.

Hauser John, Gerard J Tellis, Abbie Griffin. 2006. Research on Innovation: A Review and Agenda for Marketing Science. Marketing Science. Linthicum:. Vol. 25, Iss. 6; p. 687.

MacAfee, Andrew P., 2006. Enterprise 2.0: The Dawn of Emergent Collaboration. MIT Sloan Management Rev Spring Vol.47 No.3.

Robson David, 2007, Innovation 2.0, tapping Un-tapped talents, INNO Views Policy Workshop – Innovation Culture, Scottish Enterprise.

Rogers, E. M. 1995. Diffusion of innovation. New York: Free Press.

Schumpeter J, 1939. Business Cycles: A theoretical, historical and statistical analysis of the Capitalist process.

Schumpeter J, 1942. Capitalism, Socialism and Democracy, 3rd Edition 1950 - Harper Torchbooks, New York, 1962.

Tushman, M & Nadler, D, 1996, Organising for Innovation, California Management Review, 28 (3), 74-92.

Van de Ven, A., & Engleman, R. 2004. Central problems in managing corporate in-novation and entrepreneurship. Advances in Entrepreneurship, Firm Emergence and Growth, 7: 47–72.

von Hippel, E. 1988. The sources of innovation. Cambridge, MA: MIT.

Wheatley. M.J. ,1992. Leadership and the New Science: Learning About Organizations From an Orderly Universe. San Francisco: Berrett-Koehler.

Wolfe R, 1994, Organizational Innovation: Review, Critique and Suggested Research Directions, Journal of Management Studies 31:3.

Linking Unlearning with Innovation through Organizational Memory and Technology

Juan-Gabriel Cegarra-Navarro[1], Gabriel Cepeda-Carrion[2] and Daniel Jimenez-Jimenez[3]
[1]Universidad Politécnica de Cartagena, Spain
[2]Universidad de Sevilla, Spain
[3]Universidad de Murcia, Spain

First published in the Electronic Journal of Knowledge Management (www.ejkm.com) vol 8 issue 1, 2010.

Editorial Commentary

According to Hamel and Prahalad (1994, p. 71) "Companies are going to have to unlearn a lot of their past – and also to forget it! The future will not be an extrapolation of the past."

"Unlearning" is defined as a renewal activity to replace knowledge which becomes obsolete in turbulent environments (Hedberg , 1981). In this chapter, the concept is associated with the notion of *innovativeness* and considered as a source of organizational innovation when applied to corporate information technologies and memories which are the usual sources of corporate learning. Cegarra-Navarro, Cepeda-Carrion Jimenez-Jimenez interestingly question traditional Knowledge Management as well as the real-time relationship between technological infrastructure and corporate memory using this notion of "unlearning" so as to challenge organizational knowledge. The constant effort to unlearn enhances innovativeness by opening horizons for environmental inquiry and

the acquisition of new knowledge. This leads to dynamic organizational routines and real time adaptive behaviour for enhancing innovativeness.

References

Hamel, G. and Prahalad, C.K. (1994), Competing for the Future,
Harvard Business School Press, Boston, MA.

Hedberg, B. (1981), "How organizations learn and unlearn", in
Nystrom, P.C. and Starbuck, W.H. Eds), Handbook of Organizational Design,Vol. 1, Oxford University Press, Oxford,

Abstract: While the information technologies provide organizational members with
explicit concepts, such as writing instruction manuals, the 'organizational memory'
provides individuals with tacit knowledge, such as systematic sets, routines and
shared visions. This means that individuals within an organization learn by using
both the organizational memory and the information technologies. They interact to
reduce organizational information needs contributing to improve organizational
innovativeness. However, the utilization of the organization memory or the technology infrastructure does not guarantee that appropriate information is used in
appropriate circumstances or that information is appropriately updated. In other
words, previous memories reflect a world that is only partially understood and
assimilated, which might lead individuals to doing the wrong things right or the
right things wrong. This paper examines the relative importance and significance of
the existence of unlearning to the presence and nature of 'organizational memory
and technology'. We further examine the effect of the existence of organizational
memory and information technology on conditions that promote organizational
innovativeness. These relationships are examined through an empirical investigation of 291 large Spanish companies. Our analysis found that if the organization
considers the establishment of an unlearning culture as a prior step in the utilization of organization memory or the technology infrastructure through organizational innovativeness, then organization memory and technology have a positive
influence on the conditions that stimulate organizational innovativeness.

Keywords: unlearning, technology, organizational memory, and innovation

1. Introduction

There are two types of storage 'bins': human and no-human (Cross and
Baird, 2000). By storage 'bin', Tsang and Zabra mean "a location where
information is stored" (2008: 1444). Organizations frequently increase
their information base by using the organizational memory and the technology infrastructure. While organizational memory may be thought as

being comprised of stocks of data, information and knowledge (the memories) that have been accumulated by an organization over its history (Walsh and Ungson, 1991), the technology infrastructure represents a collection of tools for capturing and sharing of people's knowledge, promote collaboration, and provide unhindered access to an extensive range of information (Zack, 1998).

If an organization wants to start its innovation culture by bringing together these two storage 'bins': human and no-human, then it should begin to remove the obstacles that inhibit the utilization of its organizational memory and its technology infrastructure. The accuracy of that memory and the technology structures under which that knowledge is distributed and used as a constraint become crucial characteristics of organizing. While the organizational memory is comprised of all active and historical information about an organization that is worth sharing, managing and preserving for later reuse (Megill, 1997), the technology infrastructure is responsible for maintaining the networks that organizational members use to run their activities, including the data centres and software that enable the information to be used as a platform upon which the decisions are made (Gold et al., 2001).

However, there is a problem with the arguments above in that technology tools, and therefore information and organizational memory, can become obsolete at both an explicit and tacit levels. Regarding this, the negative impacts of theories in use (in terms of biases in recall, belief systems and blind spots) on decision-making have been discussed by several authors (e.g. Larwood and Whitaker, 1997). In addition, technologies can also become obsolete (Gold et al., 2001). For instance, when a company decides to come up with a new version of Windows, some old software become obsolete, and some old hardware become too inadequate to support the requirements of the new version. As Tsang and Zabra, (2008) noted, the age of an organization often gives rise to ossified routines in organizational memory.

The renewal of organizational memory or the technology infrastructure requires what Akgün et al. (2007) refer to as the 'unlearning context,' that is, the context through which the management supports the proactive questions of existing assumptions and beliefs potentially leading to being ignored, modified, deleted or replaced. In this paper we propose that the result of this unlearning context will be the updated information stored in the organizational memory and shared by using technologies. Thus, this study aims to examine the impact of an organization's 'unlearning context' on the challenging of basic beliefs or processes that are explicitly or tacitly represented in organizational memory and technology systems. We also examine the effect of the existence of organizational memory and information technology on conditions that promote organizational innovativeness. In the following we examine the concepts discussed above and explore potential relationships between them.

2. Contextual framework

An environment's discontinuities are difficult for firms to manage because they demand different product architectures, they change the economics of the industry, destroy existing firm competences, create new value networks in which to compete and require technology investments with highly uncertain outcomes (Christensen and Rosenbloom, 1995). In this context, innovation is increasingly considered to be one of the key drivers of the long-term success of a firm in today's competitive markets. The reason is that companies with the capacity to innovate will be able to respond to environmental challenges faster and better than non-innovative companies (Damanpour and Gopalakrishnan, 1998). Innovation has been conceptualised in a variety of ways in the literature, depending on the perspective from which it has been studied. It has been considered as a process; a result of both and different types of innovation have been distinguished. According to Damanpour and Gopalakrishnan (1998), innovation could be understood as the adoption of a new idea or behaviour in an organization. Literature classifies innovation between technical and administrative innovations. Whereas technical innovations include new technologies, products and services, administrative ones refer to new procedures, policies and organizational forms (Dewar and Dutton, 1986). Similarly, organizational innovativeness characterizes an organization by being supportive and permeable to innovation in terms of developing new products or processes, opening new markets, or simply developing a new strategic direction (Wang and Ahmed, 2004).

As we have discussed above, organizational memory and technology infrastructure use different retention structures. On the one hand, the most obvious structures for encoding technologies include information systems such as corporate manuals, databases, filing systems, etc (Cross and Baird,

4

2000). These systems are continually being updated and analysed and are thus capable of generating new streams of information, thereby expanding knowledge (Zuboff, 1988). On the other hand, Walsh and Ungson (1991) suggest that organizational memory is 'represented' by many diverse aspects of an organization, for example: the organization's culture, transformations (production processes and work procedures), structure (formal organizational roles), ecology (physical work settings) and information archives (both internal and external to the organization).

It is obvious that all information stored in the organizational memory or the technology infrastructure does not stay there permanently. In this regard, researchers have taken several approaches to unlearning or forgetting (Akgün et al., 2007). On one hand, in situations where organizations and their members face changing environments it is necessary that the old 'knowledge' represented in the organizational memory be challenged prior to the addition of new knowledge (Akgün et al., 2007). This idea is recognized by Huber (1996), who suggested that the basic requirement for real learning consists of abandoning manners, experience, knowledge and beliefs that are vivid and were once useful, but are not valuable in the present. On the other hand, technology infrastructure can quickly become outdated as technology, personnel and business lines change, so regularly scheduled plan maintenance and regular testing are essential to ensure team leaders are familiar with the new technology and how it relates to the company's overall business (Gold et al., 2001).

The above considerations lead us to argue that for a given organization, both the organizational memory and the technology infrastructure, needs to be critically examined since it may no longer be relevant. The unlearning context, at its heart, attempts to reorientate organizational values, norms and/or behaviours by changing cognitive structures (Nystrom and Starbuck, 1984), mental models (Day and Nedungadi, 1994), dominant logics (Bettis and Prahalad, 1995), and core assumptions which guide behaviour (Shaw and Perkins, 1991). If this is so, the context where unlearning can take place could be considered the genesis of a competitive advantage (Sinkula et al., 1997). According to Bogenrieder (2002), managers need to foster an unlearning context which opens the way for new habits, patterns, ways of doing and interpreting things to take place. To this end, Sinkula et al. (1997) propose that open-mindedness (i.e., a willingness to consider ideas and opinions that are new or different) is associated with the context of unlearning, through which the management supports the proactive questions of existing organizational routines, assumptions and beliefs potentially leading to being ignored, modified, deleted or replaced. Following Cegarra and Sanchez's (2008) suggestions, we identify the following three interaction processes that characterize an unlearning framework:

- The examination of lens fitting, which refers to an interruption of the employees' habitual, comfortable state of being, and it is through such framework that individuals of an organization will have access to new perceptions.
- The framework for changing the individual habits, which refers to the challenge of inhibiting wrong habits when an individual has not only understood the new idea, but is quite motivated to make the change.
- The framework for consolidating the emergent understandings, which refers to the organizational process that can free employees up to apply their talents by implementing new mental models based on adaptation to new knowledge structures.

Thus, we propose H_1 and H_2 based on the importance of unlearning old knowledge as a prior step to the utilisation of organizational memory and the technology infrastructure and of the negative consequences of yielding to inertial forces (Akgün et al., 2007). From this perspective, the unlearning process can be seen as the abandonment of practices that were dominant but are now standing in the way of new learning and therefore of organizational competitiveness. Therefore:

H_1: Unlearning process → Technology infrastructure

H_2: Unlearning process → Organizational memory

As noted above, an unlearning context fosters an interruption of the employees' habitual, comfortable state of being (e.g. identifying problems, initiating projects or introducing novelties). A sudden change in those habits forces individuals to reconsider their old basic attitudes toward customers, competitors, suppliers, etc. However, at this stage updated-knowledge (e.g. new meanings) is individual rather than social, and tacit rather than explicit. This knowledge then needs to be embedded through the organizational memory and the technology infrastructure in order to become a dominant design, otherwise innovation processes will not take place (Akgün et al., 2007). In this aim, new knowledge may be further 'consolidated' through the emergent understandings that are created by group members when they interact (Schein, 1992), or by new technological tools that may offer a better way to deliver information (Cross and Baird, 2000). Considering this, we argue that unlearning may have an indirect effect on innovation processes by providing support through the use of new technologies and by changing the ways individuals interact or come to interpret things. Regarding this, organizational memory and technology infrastructure have often been presented as constructs with beneficial effects on innovation processes of an organization. For example, scholars have argued that by routinizing search activities in the form of standard operating procedures, individuals can learn to become more efficient at performing them (Walsh and Dewar, 1987). Organizational memory and technology infrastructure can also provide support to individuals by retaining a broader range of potential responses, thus providing more options for organizational decision makers when they respond to the variety presented to them by changes in the organizational environment. March has asserted that 'for most purposes, good memories make good choices' (1972).

Since much of the organization's innovation is created as a consequence of the utilization of the organizational memory and the technology infrastructure interaction, it is likely to be no longer relevant due to outdated assumptions about the use of technologies. Therefore:

H_3: *Technology infrastructure* → *Innovativeness*

H_4: *Organizational memory* → *innovativeness*

3. Methodology

The population used in this study includes Spanish organizations with more than 100 employees. Like other studies on these topics, this study was designed to cover a wide range of industries (excluding the agricultural and construction sectors). 2,160 companies, from the SABI database, were located and contacted for participation. The information was collected via a postal survey directed to the R&D or innovation executive. The information-collecting period lasted from January to April 2008. The unit of analysis for this study was the company. 291 questionnaires were obtained. It is thus within the 10-20 percent average range for top-management survey response rates (Menon et al., 1996). Respondent and non-respondent companies were compared in terms of size and performance. No significant differences were found between those two groups, suggesting no response bias.

This study mainly used existing scales from literature. The questionnaire constructs comprised (see items in the Appendix):

We modelled 'unlearning context' as a formative second-order construct. We assessed 'unlearning context' by three first-order factors or dimensions: 'consolidation of emergent understandings', 'the examination of lens fitting', and 'the framework for changing individual habits'. The measures relating to the existence of a framework for 'consolidating the emergent understandings' scale consisted of 6 items taken from a scale designed by Cegarra and Sanchez (2008). To measure the framework for examining the lens fitting, 5 items were used. The final depurated scale consists of 4 items. We measured "the framework for changing individual habits" dimension through 7 items.

We adopt the formative way for our second-order construct. In this way, an increase in the level of each dimension does not imply an increase in the level of the other dimensions. The measures associated with technology are based on the infrastructure capabilities used by Gold, Malhotra, and Segars (2001). The initial scale comprises 7 items, but after the depuration process, 3 items formed this scale. Organizational Memory: We adopt a Chang and Cho (2008) scale comprises 4 items. Finally, we have used organizational innovativeness construct. Innovation has been measured in a variety of ways in previous research. In this study we measure how supportive and permeable to innovation the company is in terms of developing new products or processes. Hence, we focus on organizational innovativeness. According to Hurley and Hult (1998), innovativeness is understood as "the notion of openness to new ideas as an aspect of a firm's culture". In this paper, we measured innovativeness using a scale of 5 items adapted from Hurley and Hult. Three items make up this depurated scale.

The hypotheses were tested simultaneously using partial least squares (PLS), a structural equation modelling technique (Chin, 1998). PLS was selected due to the characteristics of our model and sample. Our model uses formative indicators and our data is non-normal. For hypothesis testing, we used the bootstrapping procedure recommended by Chin (1998). This study uses PLS-Graph software. Using PLS involves following a two-stage approach. The first step requires the assessment of the measurement model. This analysis is performed in relation to the attributes of individual item reliability, construct reliability, average variance extracted (AVE), and discriminant validity of the indicators of latent variables. For the second step, the structural model is evaluated. The objective of this is to confirm to what extent the causal relationships specified by the proposed model are consistent with the available data.

To analyse the relationships between the different constructs and their indicators, we have adopted the latent model perspective, in which the latent variable is understood to be the cause of the indicators and, therefore, we refer to reflective indicators for first-order constructs or dimensions. Three constructs in the model are operationalized as reflective, while one constructs: 'unlearning context,' is modelled as a second-order formative construct.

With regard to the measurement model, we began by assessing the individual item reliability (Table 1). The indicators exceed the accepted threshold of 0.707 for each factor loading (Carmines and Zeller, 1979). From an examination of the results shown in Table 1, we can state that all of the constructs are reliable. They present values for both Cronbach's alpha coefficient and for a composite reliability greater than the value of 0.8 for basic research (Nunnally, 1978). The AVE should be greater than 0.5, meaning that 50% or more variance of the indicators should be accounted for (Fornell and Larcker, 1981). All constructs of our model exceed this condition. We tested discriminant validity in two ways; we have compared the square root of the AVE with the correlations among constructs, and we also reported the factor scores matrix in Table 1. On average, each construct relates more strongly to its own measures than to others.

The evaluation of formative dimensions of the high-order construct: 'unlearning context', is different from that of reflective ones. One examines the weights (Mathieson et al., 2001), which represent a canonical correlation analysis and provide information about how each indicator contributes to the respective construct (see Table 2). Weights do not need to exceed any particular benchmark because a census of indicators is required for a formative specification (Diamantopoulos and Winklhofer, 2001). The concern with formative dimensions is potential multicolinearity with overlapping dimensions, which could produce unstable estimates (Mathieson et al., 2001). Results of a colinearity test show the variance inflation factor (VIF) scores of the second-order construct for all dimensions is far below the common cut-off of 10. In addition, we confirmed the validity of the formative dimensions using the procedures suggested by Fornell and Larcker (1981).

Table 1: Factor loadings of relative constructs

	Understanding	Lens	Individual Habits	Innovation	O. Memory	Technology
p1v1	0.829	0.396	0.495	0.523	0.063	0.434
p1v2	0.720	0.415	0.354	0.475	0.226	0.346
p1v3	0.828	0.180	0.314	0.537	0.087	0.227
p1v4	0.844	0.467	0.537	0.497	0.063	0.501
p1v5	0.845	0.443	0.550	0.488	0.041	0.452
p1v6	0.790	0.476	0.678	0.441	0.145	0.472
p2v1	0.387	0.791	0.383	0.317	0.135	0.295
p2v2	0.235	0.791	0.233	0.191	0.102	0.138
p2v3	0.382	0.802	0.352	0.282	0.170	0.309
p2v4	0.454	0.837	0.503	0.389	0.086	0.367
p3v1	0.501	0.460	0.822	0.363	0.171	0.436
p3v2	0.459	0.422	0.872	0.320	0.078	0.365
p3v3	0.377	0.344	0.871	0.233	0.109	0.379
p3v4	0.487	0.381	0.873	0.311	0.029	0.352
p3v5	0.361	0.354	0.845	0.263	0.130	0.217
p3v6	0.425	0.359	0.846	0.310	0.055	0.242
p3v7	0.447	0.400	0.760	0.287	0.174	0.309
p6v1	0.503	0.379	0.400	0.877	0.189	0.347
p6v2	0.507	0.375	0.388	0.887	0.259	0.338
p6v3	0.412	0.253	0.326	0.884	0.297	0.356
p12v1	0.128	0.132	0.124	0.232	0.821	0.120
p12v2	0.080	0.126	0.092	0.247	0.735	0.111
p12v3	0.077	0.097	0.087	0.222	0.715	0.107
p12v4	0.165	0.152	0.124	0.283	0.866	0.171
p14v1	0.485	0.391	0.450	0.379	0.162	0.906
p14v2	0.538	0.409	0.497	0.383	0.179	0.923
p14v3	0.376	0.291	0.363	0.288	0.039	0.831

Table 2: Weights of formative construct dimensions

High order constructs and their dimensions	weights	t de Student
Unlearning Context		
Consolidation of emergent understandings	0.55	4.49
The examination of lens fitting	0.21	2.10
The framework for changing individual habits	0.38	2.88

4. Results

The structural model resulting from the PLS analysis is summarized in Figure 1, where the explained variance of endogenous variables (R^2) and the standardized path coefficients (β) is shown. As is observed, all hypotheses presented are significant, and therefore, have been verified. Since PLS makes no distributional assumptions in its parameter estimation, traditional parameter-based techniques for significance testing and model were used (Chin, 1998). One consequence of the comparison between covariance structure analysis modelling approaches and PLS is that no proper overall goodness-of-fit measures exist for models using the latter (Hulland, 1999). The structural model is evaluated by examining the R^2 values and the size of the structural path coefficients. The stability of the estimates is examined by using the t-statistics obtained from a bootstrap test with 500 resamples. Figure 1 sets out the model statistics and the path coefficients with the level of significance achieved from the bootstrap test.

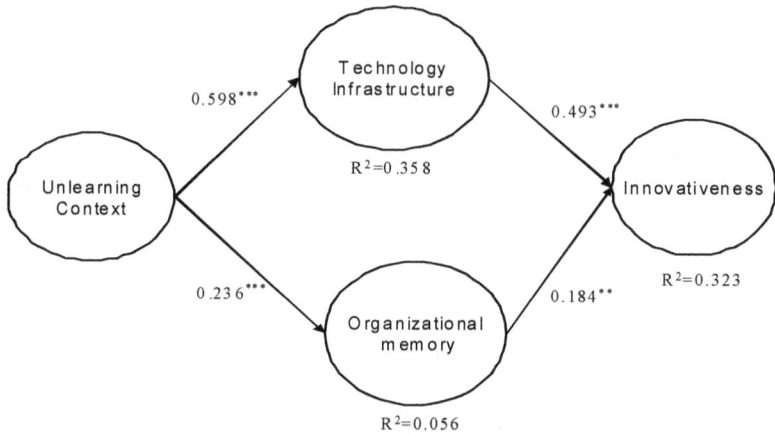

*p < 0.05; **p < 0.01; ***p < 0.001 (based on $t_{(499)}$, one-tailed test)

Figure 1: Estimated casual relationships in the structural model

Consistent with hypotheses 1 and 2, the paths between Unlearning context and Technology infrastructure (β=0.598, p<0.001) and Organizational memory (β=0.236, p<0.001) indicate positive and significant relationships among them (see Figure 1). Additionally, hypotheses 3 and 4 are supported because the two path driven to innovativeness, from technology infrastructure (β=0.493, p<0.001) and from organizational memory (β=0.184, p<0.01) are also positive and significant. What means the propensity to innovate is influenced by the technology infrastructure and the organizational memory. Finally, we performed the Stone-Geisser test of predictive relevance to assess model fit in PLS analysis (Chin, 1998). When q-square is greater than zero, the model has predictive relevance. In our model, q-square was 0.11. Regarding the effect of Technology infrastructure and Organizational memory on innovativeness, the results further suggest that the effect associated with the technology infrastructure is stronger than the effect associated with the organizational memory.

5. Discussion

This research's first contribution derives from the results of the model's empirical test. As shown in Figure 1, the results indicate that the updating of the organizational memory and the technology infrastructure depends on the unlearning context, and that 'organizational innovativeness' is unlikely on an organizational basis without being fostered by 'the organizational memory and the technology infrastructure'. A possible explanation for that could be that organizational members are unaware of the negative consequences of the outdated values, beliefs, underlying assumptions, attitudes, and behaviors shared though the organizational memory and the technology infrastructure. Under this framework, individuals become less creative and they allow individual relationships to become 'fixed' and un-responsive, thus potentially reducing the degree of innovativeness.

Therefore, the paper's first contribution stresses that companies may be over-investing in the adoption of the organizational memory and the technology infrastructure, and under investing on mechanisms to facilitate the updating of the organizational memory and the technology infrastructure.

In testing H1 and H2, our findings demonstrate a bi-directional association between the unlearning context and the technology infrastructure, and between the unlearning context and the organizational memory. Thus, potentially 'unlearning context' provides an environment that supports the modification of organizational memory when this proves necessary. These findings support the proposition by Sinkula et al. (1997) that through the creation of an 'unlearning context' in the form of 'supporting small changes, encouraging the taking of risks and cooperation' managers can change both the technology infrastructure (e.g., through justifying modifi-cations to existing technologies or even the abandonment of previously used technologies) and also organizational memory (erasing or revising older routines or organizational procedures). Therefore, the 'unlearning context' acts as "a company's capacity for organizational self-renewal and innovation through the revision of the organizational memory and the technology infrastructure in order to provide updated access to a wide range of information and knowledge". It means that through the unlearning context, organizations foster a capacity where teams and their members are continuously able to increase their abilities to articulate knowledge and use technology tools.

With regard to H3 and H4, our findings suggest that organizational innovativeness is driven by the utilization of technology and the exploitation of what has already been learned and stored into the organizational memory. The results further suggest that the effect associated with the technology infrastructure is stronger than the effect associated with the organizational memory. This result is worthy of further investigation. One conclusion that might be drawn from this result is that people usually take advantage of technologies after colleagues direct them to a specific location in a system for lessons or tools (Gold et al., 2001).

The study is not without limitations. Firstly, although in this paper, we have structured the process flow only in one direction from the unlearning context to the innovativeness, we think that there are different ways of unlearning. Under this framework, while unlearning and innovativeness could be parallel processes in a specific context, in another context the same process flow should consider unlearning as a prior step. Secondly, in the same way that each person tries to invent his or her own memorization techniques, each organization will have basic ideas on which are the capacities of each one, a circumstance which is going to facilitate the unlearning of outdated knowledge. This means that other factors, which have not been included in this study, might also affect the constructs and relationships between them. In fact, some common problems for each company would be established, which are seen and used differently by each one of them.

6. Conclusions

Based on the above arguments, the main contribution of this research is to question the existing models which relate organizational memory, technology infrastructure and organizational innovativeness. In this paper, we warn companies about the cost of outdated technology infrastructures and knowledge stored in organizational memory. Although organizational memory and technology infrastructure potentially facilitates information sharing and joint sense making if organizational memory is not updated appropriately individual learning is likely to suffer causing a reduction in the value of knowledge. Regarding this, it is often stated that firms more often converge rather than reorient, because of factors such as organizational inertia. A possible explanation for that inertia may relate to outdated memories, which can affect the learning of an individual by: narrowing the cognitive processes of individuals; hindering their ability to plan, reason and understand the situation effectively; fostering a sense of the inadequacy of linkages between variables, such as people and processes; and limiting individual's prior knowledge of the potential interactions between new processes and their consequences. This means that individuals come to rely on embedded knowledge and reduce their participation in spontaneous interactions with their colleagues. Thus individuals become less creative and they allow individual relationships to become 'fixed' and unresponsive thus potentially reducing the value of new knowledge. Therefore, when an existing stock of knowledge has already been stored by an organization, but appears counter-intuitive or deeply flawed, this current "stock" should be ignored or at least set aside temporarily if we wish to give a new idea or interpretation fair consideration. Otherwise, individuals will experience fear, pressure and uncertainty and will feel confused at the prospect of unlearning an old habit and implementing a new one, which could hinder the learning process.

Appendix 1: Questionnaire items

The consolidation of emergent understandings: with respect to your organization indicate the degree of agreement or disagreement (1= strongly disagree and 7= strongly agree):
P1_1: Managers seem to be open to new ideas and new ways of doing things
P1_2: Management has tried to initiate projects and introduce innovations
P1_3: Managers recognize the value of new information, assimilate it, and apply it
P1_4: Managers adopt the suggestions of the personnel in the form of new routines and processes
P1_5: Managers tend to collaborate with members of the organization and to solve problems together
P1_6: Managers ensure that everyone knows how to respond when faced with unexpected events
The examination of lens fitting: with respect to your current position indicate the degree of agreement or disagreement (1= strongly disagree and 7= strongly agree):
P2_1: Employees are able to identify problems (new ways of doing things) easily

P2_2: Employees are able to see mistakes from my colleagues
P2_3: Employees are able to listen to my customers (e.g. complaints, suggestions)
P2_4: Employees are able to share information with my boss easily
P2_5: Employees try reflecting on and learning from their own mistakes
The framework for changing the individual habits: with respect to your personal skills indicate the degree of agreement or disagreement (1= strongly disagree and 7= strongly agree):
P3_1: New situations have helped individuals identify their own mistakes
P3_2: New situations have helped individuals recognize undesirable attitudes
P3_3: New situations have helped individuals identify improper behaviors
P3_4: Individuals recognize their ways of reasoning or of arriving at solutions are in-adequate
P3_5: New situations have helped individuals change their behaviors
P3_6: New situations have helped individuals change their attitudes
P3_7: New situations have helped individuals change their thoughts
Technology: (1= strongly disagree and 7= strongly agree):
P14_1: There are rules for formatting or categorizing knowledge in my organization
P14_2: There are specified keywords that need to be used for categorizing or searching for knowledge in my organization
P14_3: There are common technologies available for everyone in my organization
Organizational memory: Prior to the project, my division had (1= strongly disagree and 7= strongly agree):
P12_1. A great deal of knowledge about this product category
P12_2 A great deal of experience in this product category
P12_3. A great deal of familiarity in this product category
P12_4. Invested a great deal of R&D in this product category
Organizational innovativeness: (1= strongly disagree and 7= strongly agree):
P6_1. Technical innovation, based on research results, is readily accepted
P6_2. Management actively seeks innovative ideas
P6_3. Innovation is readily accepted in program/project management
P6_4. People are penalized for new ideas that don't work (R)
P6_5. Innovation is perceived as too risky and is resisted (R)

References

Akgün, A.E., Byrne, J.C., Lynn, G.S. & Keskin, H. (2007) "Organizational unlearning as changes in beliefs and routines in organizations". *Journal of Organizational Change Management*, Vol 20, No.6, pp. 794-812.

Bettis, R.A. & Prahalad, C.K. (1995) "The dominant logic: Retrospective and extension", *Strategic Management Journal*, Vol 16, No. 1, pp. 5-14.

Bogenrieder, I. (2002) "Social architecture as a prerequisite for organizational learning", *Management Learning*, Vol 33, No. 2, pp. 197-216.

Carmines, E.G. & Zeller, R.A. (1979) *Reliability and validity assessment*. London: Sage.

Cegarra, J.G. & Sanchez, M. (2008) "Linking the individual forgetting context with customer capital from a seller's perspective", *Journal of the Operational Research Society*, Vol 59, No. 12, pp. 1614-1623.

Chang, D.R. & Cho, H. (2008) "Organizational memory influences new product success", *Journal of Business Research*, Vol 61, No. 1, pp. 13-23.

Chin, W.W. (1998) *The partial least squares approach to structural equation modeling*. G. A. Marcoulides, Ed. Mahwah, NJ: Lawrence Erlbaum Associate.

Christensen, C.M. & Rosenbloom, R.S. (1995) "Explaining the attacker's advantage: Technological paradigms, organizational dynamics, and the value network", *Research Policy*, Vol 24, No. 2, pp. 233-257.

Cross, R. & Baird, L. (2000) "Technology is not enough: improving performance by building organizational memory", *Sloan Management Review*, Vol 41, No. 3, pp. 69-78.

Damanpour, F. & Gopalakrishnan, S. (1998) "Theories of organizational structure and innovation adoption: The role of environmental change", *Journal of Engineering and Technology Management*, Vol 15, No. 1, pp. 1-24.

Day, G.S. & Nedungadi, P. (1994) "Managerial representations of competitive advantage", *Journal of Marketing*, Vol 58, No. 2, pp. 31-44.

Dewar, R.D. & Dutton, J.E. (1986) "The adoption of radical and incremental innovations: An empirical analysis", *Management Science*, Vol 32 No. 11, pp. 1422-1433.

Diamantopoulos, A. & Winklhofer, H. (2001) "Index construction with formative indicators: An alternative to scale development", *Journal of Marketing Research*, Vol.37, No 2, pp. 269–277.

Fornell, C. & Larcker, D.F. (1981) "Evaluating structural equation models with unobservable variables and measurement error", *Journal of Marketing Research*, Vol.27, No. 1, pp. 39-50.

Gold, A.H., Malhotra, A. & Segars, A.H. (2001) "Knowledge management: An organizational capabilities perspective", *Journal of Management Information Systems*, Vol 18, No. 5, pp. 185–214.

Linking Unlearning with Innovation through Organizational Memory and Technology

Huber, G.P. (1996) *Organizational learning: The contributing processes and the literature*. Organizational Learning, M.D. Cohen and L.S. Sproull, Eds. Thousand Oaks, CA: Sage.

Hulland, J. (1999) "Use of partial least squares (PLS) in strategic management research: A review of four recent studies", *Strategic Management Journal*, Vol 20, No. 2, pp. 195– 204.

Hurley, R.E. & Hult, G.T.M. (1998) "Innovation, market orientation and organizational learning: An integration and empirical examination", *Journal of Marketing*, Vol 62, No. 3, pp. 42-54.

Larwood, L. & Whitaker, W. (1997) "Managerial myopia: Self-serving biases in organizational planning", *Journal of Applied Psychology*, Vol 62, No. 2, 194-198.

March, J.G. (1972) "Model bias in social action", *Review of Educational Research*, Vol 42, No. 4, pp. 413-429.

Mathieson, K., Peacock, E. & Chin, W.W. (2001) "Extending the technology acceptance model: The influence of perceived user resources", *The Data Base for Advances in Information Systems*, Vol 32, No. 3, pp. 86-112.

Megill, K.A. (1997) The corporate memory. Information management in the Electronic Age. London: Bowker & Saur.

Menon, A., Bharadwaj, S.G. & Howell, R. (1996) "The quality and effectiveness of marketing strategy: Effects of functional and dysfunctional conflict in intra-organizational relationships", *Journal of the Academy of Marketing Science,* Vol 24, No. 4, pp. 299-313.

Nunnally, J.C. (1978) *Psychometric theory*. New York: McGraw-Hill.

Nystrom, P.C. & Starbuck, W.H. (1984) "To avoid organizational crises, unlearn", *Organizational Dynamics*, Vol 12, No. 4, pp. 53-65.

Schein, E.H. (1992) *Organizational culture and leadership*, San Francisco: Jossey-Bass.

Shaw, R.B. & Perkins, D.N. (1991) "Teaching organizations to learn", *Organization Development Journal*, Vol 9, No. 4, pp. 1-12.

Sinkula, J.M., Baker, W.E. & Noordewier, T. (1997) "A framework for market-based organizational learning: Linking values, knowledge and behaviour", *Journal of the Academy of Marketing Science*, Vol 25, No 4, 305-318.

Tsang, E. & Zabra, S. (2008) "Organizational unlearning", *Human Relations*, Vol 61, No. 10, pp. 1435-1462.

Walsh, J.P. & Dewar, R.D. (1987) "Formalization and the organizational life cycle", *Journal of Management Studies*, Vol 24, No. 3, pp. 216-231.

Walsh, J.P. & Ungson, G.R. (1991) "Organizational memory", *Academy of Management Review*, Vol 16, No. 1,pp. 57-91.

Wang, C.L. & Ahmed, P.K. (2004) "The development and validation of the organisational innovativeness construct using confirmatory factor analysis", *European Journal of Innovation Management*, Vol 7, No. 4, pp. 303-131.

Zack, M.H. (1998) "An MIS Course Integrating Information Technology and Organizational Issues", *Data Base*, Vol. 29, No. 2, pp. 73-87.

Zuboff, S. (1988) *In the age of the smart machine*. New York: Basic Books.

Radical innovation versus transformative learning: A Kuhnian reading

Alexandros Kakouris

Career Office and Faculty of Informatics and Telecommunications, National and Kapodistrian University of Athens, Greece

First published in The Proceedings of ECIE 2010

Editorial Commentary

As claimed in the preamble of this chapter, Kuhn's theory of "paradigm shift" and "inevitable discontinuities " is one of the most influential contribution in 20th century philosophy of science. This theory argues that crises due to the rise of anomalies in a paradigm of thought are resolvable by the adoption of a new paradigm with the espousal of new beliefs and values. The powerful emergence of ICT associated with the subsequent shift of focus of innovation has provoked a paradigm shift concerning the concept of innovation itself. In this type of complex environment, entrepreneurial learning is assigned new imperatives and innovative perspectives.

Kakouris offers a clear perspective on radical innovation and transformative learning which, together, provide real benefits to innovative entrepreneurship education. An instructive deconstruction of both concepts from a Khunian perspective is explored, and a reflection on the catalytic power of discontinuities sheds new light on the notion of "radical innovation". The study provides a rich theoretical investigation of life-long learning.

Abstract: It is widely accepted in literature that radical innovation challenges contemporary industries. Given the spectacular scientific achievements of the twentieth century, the potential of radical innovation inspires entrepreneurs in new product, services or processes development. Modern innovation has become more complex since it embodies marketing and organizational innovations in its original technological content. Notably, the definition of innovation has been broadened in the latest Oslo manual in order to meet and measure the way that modern enterprises or individuals innovate. At the same time, innovative entrepreneurship education and innovation management expands in both formal and informal learning. It is well documented by agencies in governmental reports that the impact of entrepreneurial courses depends on new and innovative teaching methods introduced in courses. Since attendees of such courses have personal assumptions or beliefs about the complex phenomenon of business venturing, transformative learning offers an efficient, yet poorly exploited, approach towards introducing innovative entrepreneurship in a classroom setting. This paper aims to discuss the two different concepts of radical innovation and transformative learning under the same Kuhnian perspective. We argue that despite their unrelated nature, both concepts involve a Kuhnian core-process based on "discontinuities", similar to epistemological paradigm shifts. Key similarities, differences and difficulties are discussed while the innovator, or the innovating organization, is contrasted to the educator that follows transformative methods. Furthermore, the relevance of different aspects or types of innovation and those of entrepreneurial learning to radical innovation and transformative learning are examined. Implications in entrepreneurship education of a possible common understanding are presented in order to suggest embedment of educational elements in teaching that trigger transformations and thus, openness of participants in entrepreneurial learning.

Keywords: radical innovation, transformative learning, paradigm shift, Kuhnian theory, entrepreneurial education

1. Introduction

Kuhn's (1962) original work for the evolution of science through revolutions, elaborated in his book entitled *"The structure of scientific revolutions"*, has been one of the most influential 20th–century essays in history and philosophy of science. Kuhnian theory implies inevitable discontinuities, i.e. revolutions, once anomalies appear in a paradigm, i.e. a consensual scientific perception, that lead to a crisis resolvable only by adoption of a new paradigm. The *paradigm shift* cannot be a cumulative process in its own time despite its usually gradual presentation in the post–revolution period. Early Schumpeterian ideas on innovation, expressed more than thirty years earlier (Schumpeter 1934), are described very much alike paradigm shifts. Such ventures, emerging from technological breakthroughs, constitute *radical* innovations. Thus, radical innovation is often presented in Kuhnian terms of paradigm shift in literature.

In the study of entrepreneurship, Cope and Watts (2000) explore the process of entrepreneurial learning in personal and business development. The authors examined the outcomes of critical events (or episodes) as powerful sources of learning experiences. The appearance of such episodes in narratives has also been pointed out by Rae and Carswell (2000, 2001). Cope (2003, 2005) discusses the meta–cognitive nature of learning that emanates from critical events since it affects *"habits of mind"* or *"beliefs/points of view"*, both terms introduced by Mezirow (1978) in his adult education theory of transformative learning. As an entrepreneur learns from experience (cf. Minniti & Bygrave 2001, Politis 2005), critical events stimulate reflection, i.e. thinking on own thoughts, similar to the double-loop procedure in organizational learning (Argyris and Schön 1978) and to the way that practitioners operate and learn (Schön 1983). Thus, Cope and Watts (2000) suggest nature of critical events be further studied and transformative learning be exploited in the study of entrepreneurship.

Meanwhile, diffusion of European entrepreneurship education, followed the Oslo agenda (2006), encourages innovative teaching in the field. Corresponding courses are found interdisciplinary in almost 50 percent of European higher institutes. Most of them are short-term or parts of other courses (European Commission 2008). Given the particularities of the subject (e.g. Fiet 2000, Pittaway and Cope 2007), courses' impact on graduates is closely related to innovative and effective teaching. For a further development of this type of education, a necessity for trainers' training employment has been discussed in Kakouris (2007). Not limited to trainers, transformative learning methods may enrich the experiential character of modules as they primarily focus on participants' beliefs. For instance, Shane (2008) addresses several such assumptions about entrepreneurship which contradict to facts or data. Concerning Greek students, Agapitou et al. (2010) present empirical results on underlying beliefs about entrepreneurship in the present volume. Expedient teaching should deal with attendees' beliefs and we argue that, to our knowledge, transformative learning has not been adequately exploited in that direction so far.

The present conceptual article aims to juxtapose radical innovation and transformative learning from a Kuhnian perspective. The Kuhnian character of the former was already mentioned while Mezirow refers, in his introductory note of *"Learning as transformation"* (Mezirow and associates 2000), an early influence from the concept of Kuhnian paradigm. Such a comparison especially aims to contribute to a common comprehension and adoption of transformative learning in entrepreneurial education. Furthermore, it attempts for a pursuit of radicalism in innovation as the recent meaning of the term has been extended (OECD 2005) to increase its humanistic content compared to the technological one.

2. Concepts of discontinuous innovation and learning

2.1. Radical innovation

Innovation as a means of economic development and growth appeared in the work of Schumpeter (1934). The Schumpeterian innovation is radical since it displaces the market equilibrium under a *creative destruction* mechanism. Radical innovation may also open new markets and is accompanied by rapid economic growth. It was initially attributed to technology breakthroughs emerging from applied scientific research in accordance to a linear model of innovation (Godin 2006). Radical innovation is unpredictable and exhibits an inherent "human side" (O'Connor and McDermott 2004) that concerns either the innovator, i.e. the founder of a new firm, or potential employees in large firms that adopt innovation management. Innovation, either product/service or process, has been traditionally assumed as a matter of technology. However, the latest definition in the Oslo manual (OECD 2005) incorporates marketing innovation and organizational innovation; both closely related to human capital. Therefore, more practically and especially for measurement purposes, radical innovation can be defined as: *"a project with the potential to produce one or more of the following: an entirely new set of performance features; improvements in known performance features of five times or greater; or a significant (30 percent or greater) reduction in cost"* (Leifer et al. 2000: 5). Compared to the outcomes of the great inventions of 20th century, there is need for radical innovations in contemporary business venturing. However, large companies exhibit resistance in following proper opportunities due to the high risk associated with radical innovation.

Limiting the discussion to technological innovation, radical innovation is also referred as "discontinuous" contrary to the incremental one. Similar to its progenitor (i.e. scientific research) discontinuities in technology appear during scientific revolutions as described by Kuhn (1962). A customary schematic view of a radical innovation is illustrated in Figure 1. The grey plot shows the evolvement of technology in accordance to the Kuhnian structure of scientific revolutions (SSR). "Normal" technology (coming from the Kuhnian normal science) evolves incrementally before and after a crisis period where the old paradigm is suddenly replaced by the new one, denoted by the "jump" in the diagram. The black plot illustrates the curve of a product that is based on the specific technology. Product curves (1 and 2) are sigmoid due to diffusion of innovation (Rogers 1962). In the appearance of the new technology (i.e. the new paradigm) the product manufacturer has to consider the adoption of radical innovation. Such an adoption will change both the technology of the product and the business model of the firm. However, an adherence in the old technology will gradually lead to a sales decay, i.e. curve 1 will continue in dotted curve 3. Under the adoption of innovation, curve 2 initiates a new business cycle for the advanced product. The way that curves 1 and 2 coexist during the "technology–shift" period is not unique. It depends on the decisions of the manufacturing firm or its competitors who will exploit the new technology. Evidently, the transition from curve 1 to 2 is considered discontinuous. Innovation during phases 1 and 2 is incremental.

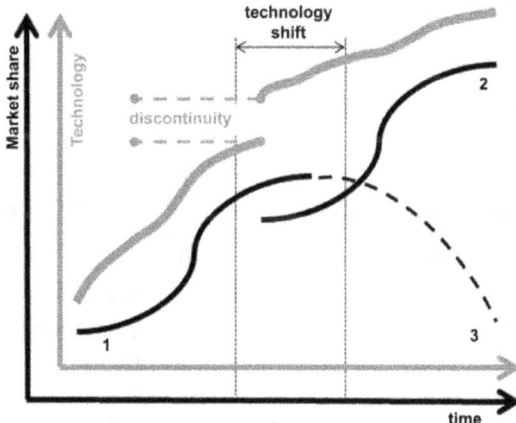

Figure 1: A schematic representation of radical innovation

Critique on the emergence of radical innovation in the market and the feasibility of a plot similar to Figure 1, has been provided by Rosenberg (1976).

The author argues that there is lack of examples of radical Schumpeterian innovation since new technologies enter the market gradually due to commercialization procedures. He also notes that even for exceptionally radical new combinations, the propagation in market demands progressive modification and adaptation to suit submarkets. Studying the innovation process in microcomputers, Bhidé (2000) provides additional evidence for gradual deployment of innovation. The author explains the way that small and large firms innovate in a complementary way. Initially, the innovation originates stochastic with many small innovative firms competing in the new technology. In a later stage, as some firms grow or as large firms adopt the innovation it becomes strategic focusing on the long–term outcomes. Because large firms avoid "cannibalizing" their existing customers they insert new technologies gradually in the market and in a more predictable way. Inadequate financing of a new technology can also be another restrictive factor for its fast diffusion in the market. Thus, the emergence of discontinuous innovation is questionable in accordance to market observations. Kuhn (1962: 138) addresses the historical presentation of a paradigm shift in the post–revolution phase. The argument pertains to the rewriting of history by authorities in a linear manner which follows the pedagogy of textbooks. He notably refers in chapter XI of his essay entitled *"The invisibility of revolutions"*: *"... No wonder that textbooks and the historical tradition they imply have to be rewritten after each scientific revolution. And no wondered that, as they are rewritten, science once again comes to seem largely cumulative..."*. In conclusion, radical innovation in technological context can be considered inherently discontinuous, however, it needs thorough research to be identified.

2.2. Transformative learning (Mezirow 1978, 1991, 2000)

The way that an entity (person or organization) learns is far from being clear. Due to a holistic consideration of learning, a human response on facts and external evidence is more educationally consistent and reliable than the organizational one which may suffer from misconceptions of internal roles of groups, poor collaboration, deficient communication, etc. But even ordinary learning can be either formative or transformative. The former is considered continuous (incremental) coming from formal sources in contrast to the latter that includes "discontinuities" during critical events or "episodes". In adulthood, the person possesses an established frame of reference that not only includes knowledge but also beliefs (or assumptions), values, attitudes and feelings. Due to his tend for self-awareness and efficient response in problem-solving, the adult person is able to (re)consider his own beliefs critically; a process known as *critical self-reflection*. Mezirow introduced transformative learning in a constructivist ground, i.e. learning corresponds with meaning–making of an individual (for reviews see Taylor 2007, Kitchenham 2008). He suggests that a frame of reference includes both "habits of mind" (meaning–making) and "points of view" (beliefs). From the view of psychology, Kegan (2000) discusses Mezirow's transformation as an epistemological change. The frame of reference that undergoes transformation not only changes the formation of meaning but also reforms the meaning–forming. The author concludes that the later meta–process may be further viewed under the constructive developmental theory. Apparently, transformative learning is a higher order process to be further deployed.

The question we raise in the present work is *"how consistent with the pattern of Figure 1 is transformative learning?"*. Transformation begins with a personal "disordering dilemma" which is an unstructured event dominated by emotions. Such events, often traumatic, were described by Cope and Watts (2000) for the entrepreneurs they interviewed. The dichotomy that a person confronts during critical events echoes the interpretation of Figure 1. The transition from one frame of reference to another is not at all an incremental process since the new frame is incompatible to its pre–existing one. Transformation is also an irreversible process that dramatically changes the inherent meaning–structure (the "epistemology") of the person accompanied with large scale changes in his behavior and motivation. Besides, a mandatory and also confirmatory consequence of a transformation is action undertaking, i.e. transformative learning is not a passive process. However, it would be interesting to examine how people describe such transformations, or critical events, after a period. It is possible for them to eliminate annoying details from their memories in a way similar to Kuhn's description for post-revolution narrations of paradigm shifts. In such a case, the transformation may look like cumulative in later stages, however, further research is needed towards that direction. Conclusively, we consider transformative learning as a discontinuous process viable during unstructured critical events.

In order to employ transformative learning educational techniques, a broad definition of entrepreneurship is needed. It should not be considered as a mere business start-up but as an individual's mindset able to overcome uncertainties, cope with fear factor and creatively transform opportunities it into promising enterprises. Hence, innovative entrepreneurship is a field that can be facilitated by transformative learning.

3. Innovation and learning under a Kuhnian framework

The brief presentation of radical innovation and transformative learning in the preceding section revealed two certain commonalities between the two concepts: undertaking action (moving forward) to the "unknown" and fear tolerance. Hence, resistance to such changes is always present, for various specific reasons, and has been clearly annotated by Schumpeter, Kuhn and Mezirow. The inherent irreversibility of both processes claims robust decision making and self–confidence related to the individuals' personal attributes and idiosyncrasies. Noteworthily, descriptions of scientific paradigm shifts, radical innovations and transformative learning focus on human factor. Successful innovators become famous and role models whilst the situation in innovating organizations is more complicated.

As a corollary to the analysis so far, the Kuhnian core of radical innovation and transformative learning has been disclosed. A metaphorical match of Kuhnian paradigm with technology and frame of reference enables the common juxtaposition of five phases, shown in Table 1, in which each individual process can be decomposed.

The pre–paradigm phase in Kuhnian approach is dominated by competing, incomplete and incompatible scientific theories with no general consensus. This phase can be parallelized with the innovative business idea elaboration and learning in childhood respectively. The innovator gathers information and data relative to a specific opportunity without a predominant decision about the technology and business model he will use. Similarly, children learn from various formal sources of authority without a fully developed and consistent set of beliefs, values and attitudes. As they grow to adults, they shape own frames of reference and pursue formative learning. This is an incremental process where meaning is formed under their habits of mind while their points of view reflect their own assumptions. This is the second phase of Table 1 corresponding to Kuhnian scientific paradigm which allows cumulative normal science to develop. In the same phase, innovation is incremental as shown by curve (1) in Figure 1 and utilizes both a specific technology and a concrete business model for the innovating firm. A certain commonality during this phase is the clarity of paradigm / technology / frame and the rigid rules that imposes on problem solving / business model / learning.

Table 1: Kuhnian juxtaposition of radical innovation and transformative learning

Phase	Radical innovation interpretation	Kuhnian theory of SSR	Transformative learning theory
I	*Business idea elaboration*	*Pre-paradigm phase*	*Learning in childhood*
II	*Old technology and business model*	*Old paradigm phase* – normal science	*Old frame of reference* – well established "habits of mind" and beliefs – formative learning
III	*Old technology and business model inefficiencies* – emergence of new technology	*Crisis or "anomaly" phase* – unsolved problems – appearance and competition of new theories	*Critical event or "episode"* – problematic situation – disordering dilemma
IV	Adoption of both: – new technology and – new business model	*Paradigm shift* – Scientific revolution	*Critical reflection and/or self–reflection* – transformation (reframing) – transformative learning
V	*New technology and model* – new business cycle – new markets	*New paradigm phase* – post revolution phase – normal science	*New frame of reference* – formative learning

According to Kuhn, a phase of crisis (III in Table 1) emerges when irresolvable problems stand for a long time. In such a case the paradigm blurs, new competitive theories appear whilst the imposed rules in problem solving loosen. Accordingly, a critical event creates a disordering dilemma when an individual has to practically respond in a problematic situation. As meaning becomes less clear, a possible solution requires holistic consideration of the underlying frame of reference. The same critical phase for radical innovation pertains to its driver. The emergence of a new, incompatible and advanced technology enforces existing firms to innovate. This is a supply-driven approach whilst a demand-driven possibility is not ruled out whereas specific customer needs stand for long time creating a promising potential market. A response to crisis (phase IV) comes through the paradigm shift. The old paradigm is not rejected until replaced by the new one. The new paradigm is incompatible with the previous, solves the standing problem(s) and poses novel predictions. Discontinuity lies in phase (IV) in which cumulative progress arising from the old paradigm is not possible. Kuhn describes the transition as a gestalt switch or as a Darwinian natural choice. Transformative learning also refers to phase (IV). The individual employs critical self-reflection, transforms his frame of reference and proceeds to action through the new meaning that emerges from his transformed frame. Solution to the problematic situation that created the crisis is then possible through reframing. Finally, the adoption of radical innovation proceeds in a similar manner. Such an adoption demands simultaneous alteration of technology and business model of the enterprise. Evidently, it is a costly and uncertain procedure that probably inspires individual entrepreneurs but also puzzles large companies.

Finally, phase (V) pertains to the post–revolution situation which is analogous to phase (II). Normal science proceeds under the new paradigm, formative learning evolves in accordance to the transformed frame of reference and the innovative firm enters a new business cycle provoking the existing and possible new markets. A further discussion may concern the remark of Kuhn that "... *Revolution concerns those who are affected; outsiders see it as a normal developmental process...*" and study the way that innovators or adult learners present their past experience about the phase of crisis. This may lead to a better identification, isolation and quantification of such phenomena.

4. Conclusions

Present work discussed the different but "*Kuhnianly*" correspondent concepts of radical innovation and transformative learning. Under a metaphorical matching of *technology – paradigm – frame of reference*, five discrete phases of *radical innovation – scientific – learning* evolution were illustrated in common. This is not a new result since technology results from applied scientific research (see for instance Wonglimpiyarat 2010) whilst transformative learning theory has been influenced in its early stage by the Kuhnian perspective (Mezirow and associates 2000, Kitchenham 2008). However, the present pattern may facilitate both educators and practitioners to envisage either advanced learning methods or contemporary topics to be taught in order to enhance innovative entrepreneurship education.

In the study of entrepreneurship, Cope (2005) suggests the adoption of a "learning lens" is needed. Such a "lens" needs to include the theory of transformative learning since critical events of entrepreneurs constitute particularly rich sources for experiential learning. The present approach aims to intimate this type of learning in an articulate way to researchers familiar to innovation and science. Furthermore, the concept of innovation (OECD 2005) has become more humanistic including its marketing and organizational sub–types. Thus, radical or discontinuous innovation has to be reconsidered in a less technological context. But even discussed as pure technological, radical innovation may remain obscure for a series of reasons (cf. Bhidé 2000) that are beyond the scope of this article. Thus, innovation measurements have become more complicated and need clarifying definitions and rigorous research in order to be accurate. We believe that the present work contributes in this direction.

Finally, innovative entrepreneurship education is based on experiential learning (Kolb 1984). Relevant courses exploit learning by doing focused on business planning (e.g. Garavan and O'Cinneide 1994a,b, Hynes 1996, Honig, 2004, Rasmussen and Sørheim 2006, Smith et al. 2006). Transformative learning offers techniques that can enrich critical reflection and deal with socio–cultural beliefs or assumptions of graduates in order to advance the content and the impact of the relevant, short–term courses. Such techniques are poorly exploited so far and they will be especially important in the provision of entrepreneurial training through lifelong learning.

Acknowledgements

This work was financially supported by the Greek Ministry of Education and Religious Affairs through the "Education and Lifelong Learning" programme.

References

Agapitou, C., Tampouri, S., Bouchoris, P., Georgopoulos, N. and Kakouris, A. (2010) "Exploring underlying beliefs on youth entrepreneurship of higher education graduates in Greece", Paper read at 5th European Conference on Innovation and Entrepreneurship, Academic Conferences Limited., Reading, UK, September.

Argyris, C. and Schön, D. (1978) *Organisational Learning: A Theory of Action Perspective*, Addison Wesley, Reading.

Bhidé, A. (2000) *The Origin and Evolution of New Businesses*, Oxford University Press, Oxford.

Cope, J. (2003) "Entrepreneurial learning and critical reflection: Discontinuous events as triggers for higher level learning", *Management Learning*, Vol. 34, No. 4, pp 429–450.

Cope, J. (2005) "Toward a Dynamic Learning Perspective of Entrepreneurship", *Entrepreneurship Theory and Practice*, Vol. 29, No. 4, pp 373–397.

Cope, J. and Watts, G. (2000), "Learning by doing: An exploration of experience, critical incidents and reflection in entrepreneurial learning", *International Journal of Entrepreneurial Behaviour and Research*, Vol. 6, No. 3, pp 104–124.

European Commission (2008) "Survey of Entrepreneurship Education in Higher Education in Europe", [online], NIRAS Consultants, FORA, ECON Pöyry, http://ec.europa.eu/enterprise/entrepreneurship/support_measures/training_education/highedsurvey.pdf

Fiet, J.O. (2000) "The theoretical side of teaching entrepreneurship", *Journal of Business Venturing*, Vol. 16, pp 1–24.

Garavan,T.N. and O'Cinneide, B. (1994a) "Entrepreneurship education and training programmes: a review and evaluation - Part 1", *Journal of European Industrial Training*, Vol. 18, No. 8, pp 3–12.

Garavan,T.N. and O'Cinneide, B. (1994b) "Entrepreneurship education and training programmes: a review and evaluation - Part 2", *Journal of European Industrial Training*, Vol. 18, No. 11, pp 13–21.

Godin, B. (2006) "The Linear Model of Innovation: The Historical Construction of an Analytical Framework", *Science Technology Human Values*, Vol. 31, No. 6, pp 639–667.

Honig, B. (2004) "Entrepreneurship Education: Toward a Model of Contingency-Based Business Planning", *Academy of Management Learning and Education*, Vol. 3, No. 3, pp 258–273.

Hynes, B. (1996) "Entrepreneurship education and training – introducing entrepreneurship into non-business disciplines", *Journal of European Industrial Training*, Vol. 20, No. 8, pp 10–17.

Kakouris, A. (2007) "On a Distance-Learning Approach on 'Train the Trainers' in Entrepreneurial Education in Greece. Theoretical Considerations Supported by Students' Response as Observed from the Career Office of the University of

Athens", Paper read at 2nd European Conference on Entrepreneurship and Innovation, Academic Conferences Ltd., Reading, UK, November, pp 75–79.

Kegan, R. (2000) "What "form" transforms? A constructive-developmental approach to transformative learning", in J. Mezirow & Associates (Eds.), *Learning as transformation: Critical perspectives on a theory in progress* (pp 35-69), Jossey-Bass, San Francisco.

Kitchenham, A. (2008) "The evolution of Jack Mezirow's transformative learning theory", Journal of Transformative Education, Vol. 6, No. 2, pp 104–123.

Kolb, D.A. (1984) Experiential learning experience as a source of learning and development, Prentice Hall, New Jersey.

Kuhn, T.S. (1962) *The Structure of Scientific Revolutions*, 3rd edition (1996), University of Chicago Press, Chicago.

Leifer, R., McDermott, C.M., O'Connor, G.C., Peters, L., Rice, M. and Veryzer, R.W. (2000) *Radical Innovation: How Mature Companies Can Outsmart Upstarts*, Harvard Business School Press, Boston, MA.

Mezirow, J. (1978) "Perspective Transformation", *Adult Education Quarterly*, Vol. 28, No. 2, pp 100–110.

Mezirow, J. (1991) *Transformative dimensions in adult learning*, Jossey-Bass, San Francisco.

Mezirow, J. and associates (2000) Learning as Transformation: Critical Perspectives on a Theory in Progress, Jossey-Bass, San Francisco.

Minniti, M. and Bygrave, W. (2001) "A dynamic model of entrepreneurial learning", *Entrepreneurship Theory and Practice*, Vol. 25, No. 3, pp 5–16.

O'Connor, C.G., and McDermott, C.M. (2004) "The human side of radical innovation", *Journal of Engineering and Technology Management*, Vol. 21, pp 11–30.

OECD (2005) "Oslo Manual. Guidelines for Collecting and Interpreting Innovation Data", 3rd Edition, [online], OECD and Eurostat, http://browse.oecdbookshop.org/oecd/pdfs/browseit/9205111E.PDF

Oslo agenda (2006) "Oslo Agenda for Entrepreneurship Education in Europe", [online], European Commission, http://ec.europa.eu/enterprise/entrepreneurship/support_measures/training_ education/doc/oslo_agenda_final.pdf

Pittaway, L. and Cope, J. (2007) "Entrepreneurship education: A systematic review of the evidence", *International Small Business Journal*, Vol. 25, No. 5, pp 479–510.

Politis, D. (2005) "The Process of entrepreneurial learning: a conceptual framework", *Entrepreneurship Theory and Practice*, Vol. 29, No. 4, pp 399–424.

Rae, D. and Carswell, M. (2000) "Using a life-story approach in researching entrepreneurial learning: the development of a conceptual model and its implications in the design of learning experiences", *Education and Training*, Vol. 42, No. 4/5, pp 220–227.

Rae, D. and Carswell, M. (2001) "Towards a conceptual understanding of entrepreneurial learning". *Journal of Small Business and Enterprise Development*, Vol. 8, No. 2, pp 150–158.

Rasmussen, E.A. and Sørheim, R. (2006) "Action-based entrepreneurship education", *Technovation*, Vol. 26, pp 185–194.

Rogers, E.M. (1962) *Diffusion of innovations*, 4th edition (1995), Free Press, New York.

Rosenberg, N. (1976) *Perspectives on Technology*, Cambridge University Press, Cambridge.

Schön, D. (1983) *The reflective practitioner*, Basic Books, New York.

Schumpeter, J. (1934) *The Theory of Economic Development*, Harvard University Press, Cambridge, MA.

Shane, S. (2008) *The illusions of entrepreneurship*, Yale University Press, New Heaven.

Smith, A.J., Collins, L.A. and Hannon, P.D. (2006) "Embedding new entrepreneurship programmes in UK higher education institutions. Challenges and considerations", *Education and Training*, Vol. 48, No. 8/9, pp 555–567.

Taylor, E.W. (2007) "An update of transformative learning theory: a critical review of the empirical research (1999-2005)", *International Journal of Lifelong Education*, Vol. 26, pp 173–191.

Wonglimpiyarat, J. (2010) "Technological change of the energy innovation system: From oil-based to bio-based energy", *Applied Energy*, Vol. 87, No. 3, pp 749–755.

Strategy and Innovation: Making the Right Strategic Decision and Developing the Right Innovative Capabilities

Lawrence J. Loughnane
CETYS Universidad Mexicali, Mexico

First published in The Proceedings of ECIE 2007

Editorial Commentary

Innovation is the driving force of today's economy and an impera-tive for corporate performance and survival. Innovation is complex, uncertain and, according to Kidd et. al. (2005) almost impossible to manage ... although it is defined as, "...a core process concerned with renewing what the organization offers and the ways in which it generates and delivers these." (Tidd et. al., 2005: 40). If innovation is a process or management practice, managers can actively deal with it to plan, manage, control ... according to their own context (de Waal et. al., 2010).

Innovation as a capability and a management practice, based on strategic decisions and business vision, is the topic explored by Loughnane. According to his work each organization requires a con-sistent and contextually-designed model of what innovation is. The right management skills to create necessary alignments with pri-mary management practices (strategy, execution, culture, and structure) are mandatory. This chapter draws on the literature's major concepts to place innovation as an organizational capability that is generated from direct and appropriate organizational in-vestments.

References

Anton de Waal, Alex Maritz, Chich Jen Shieh, 2010. Managing Innovation: A Typology Of Theories And Practice-Based Implications For New Zealand Firms. International Journal of Organizational Innovation (Online). Hobe Sound: Fall 2010. Vol. , Iss. 2; p. 35

Tidd, J., Bessant, J., and Pavitt, K. (2005). Managing Innovation - Integrating technological, market and organizational change (3rd ed.). Australia: John Wiley and Sons

Abstract: All organizations must be innovative. Is this statement true or is innovation only another management practice that a company can employ in a search for superior performance? This paper suggests that innovation is a management practice and it is critical for an organization to decide if it is going to pursue innovation as a management practice. Despite the call for companies to be innovative, research indicates that it is not necessary for an organization to be innovative to be highly successful. Worse, it is problematic that few companies are capable of excelling at innovation. This is because it is a very difficult management practice and few senior leaders have a clear, well-developed model of what innovation looks like as an organizational capability within a company's specific context.

To examine the need for innovation and a company's capability to excel at innovation perhaps one starting point is to define innovation differently. There are two types of innovation: *upstream and downstream* innovation. *Upstream* innovation is development of new inventions and technologies. *Downstream* innovation is the process of turning the inventions and processes into economic value.

A company's vision drives the process of deciding to pursue innovation. Company leaders should be aware innovation exists in a context approaching chaos. It exists in a context where complexity is high and the unpredictable occurs far more frequently than predictable results.

Determining what management practices will lead a company to superior performance is the first step to becoming a leader in an industry. Companies that outperform their industry peers excel at what are called primary management practices—strategy, execution, culture, and structure. Companies that out-perform their industry peers supplement their skill in primary areas with a mastery of any *two* out of *four* secondary management practices—talent, *innovation*, leadership, and mergers and partnerships.

A company that chooses to pursue innovation, as a management practice, must first recognize that being innovative is not a strategy. Strategy is also a management practice. Being innovative is a capability that is the result of a successful strategy. A company must choose to be very good at strategy and other management practices and the practices must be aligned and performed at the same time.

Keywords: Innovation, entrepreneurship, creativity, business practice, strategy

1. Introduction

There is a constant theme in the business press and respected business publications that all organizations must be creative and/or innovative. The definitions of creation and innovation are very close; close enough that many people consider them synonyms. They are not. Creativity is about idea generation. Innovation is about idea implementation. Innovation is typically understood as the *introduction* of something *new* and *useful*, for example introducing new methods, techniques, or practices or new or altered products and services.

It is critical for an organization to decide if it is going to pursue innovation as a management practice. To be considered innovative an organization needs to change its industry in some way (Joyce, Nitin and Roberson (2003). Innovators are focused on finding altogether new product ideas or technological breakthroughs that have the potential to transform industries.

It is not necessary for an organization to be innovative to be highly successful (Joyce, et al, 2003). Once an organization decides that innovation will be a management practice, it must decide if it is going to pursue *upstream innovation* or *downstream innovation*. Upstream innovation is the implementation of one's own ideas. Downstream innovation is implementing the ideas of others. Both upstream and downstream innovations have the same goal - make investments and manage those investments so the return on the investment contributes to superior performance. Superior performance is defined as the sustained creation of value for which a customer will pay.

It is problematic that few companies are capable of excelling at innovation – it is a very difficult management practice (Hamel, 2003). Without other complimentary *management practice skills* innovation rarely, if ever, results in the creation of value. Determining what management practices will lead to superior performance is the first step to becoming a leader in an industry.

Traditionally companies have been told that industry forces (Porter, 1980) and how an organization reacts to those forces determine performance. Industry leaders, rather than focusing on external forces focus on developing internal capabilities (Hawawini, Subramanian, Verdin, 2002). It is important to note that it is the industry leaders or those that want to be industry leaders that focus on internal capabilities. Average or middle of the road companies focus on the external forces and how their strengths and weakness match those forces. To be innovative and thus be in a position to change its industry in some way requires that an organization develop the capability to be innovative as a management practice.

Joyce et al (2003) document the results of research that indicates:

- Companies that out-perform their industry peers excel at what are called primary management practices—strategy, execution, culture, and structure.
- Companies that out- perform their industry peers supplement their skill in primary areas with a mastery of any *two* out of *four* secondary management practices—talent, innovation, leadership, and mergers and partnerships.

Note that companies that exhibit superior performance excel at all of the primary management practices. But, it is not necessary for a superior performer to be innovative. Innovation is not a primary management practice. Innovation is a secondary management practice. (Secondary in this sense means that there is a decision about which two of the four secondary practices a company will use). An organization must make a decision to invest in the development of a capability to be innovative. Based on the amount of literature on innovation, it might be expected that all companies should invest in the creation of an innovative culture. The truth is few companies really make this investment. Those companies that do make the investment often fail to achieve intended objectives because it is a very difficult management practice. And, it is not always the best investment an organization can make.

2. Management Practice

Management practice can be viewed as the sides of a triangle. Each side supports the other two. The sides of the triangle are *content, context* and *process* (See Figure 1). *Content* is limited by an organization's *capability and capacity.* The *competitive environment* (industry forces) defines con-text. *Process* is characterized by the *patterns of organizational behaviour* an organization exhibits as it gets things done.

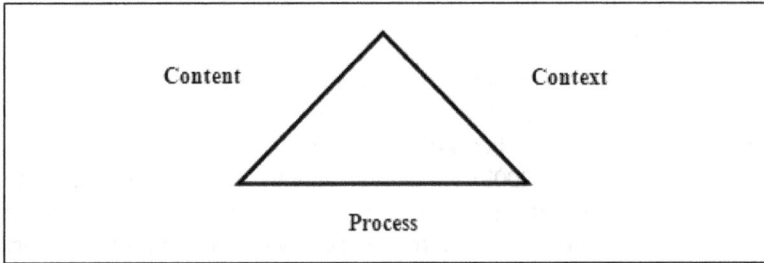

Figure 1: Management Practice (Loughnane 2007)

Management practice is contextual simply because what works in one situation (context) can easily fail in another. When trying to determine what kind of management practice will work in a particular context a good place to start is to think about context existing on a continuum (See figure 2).

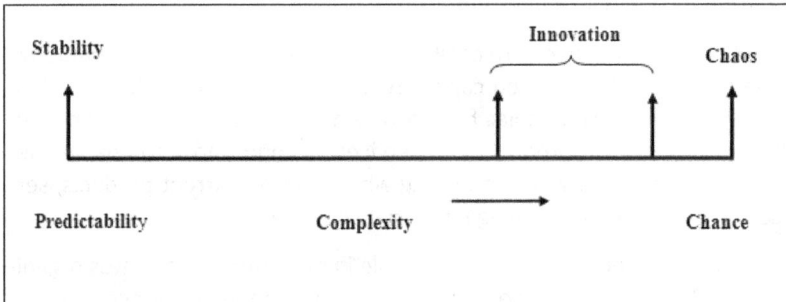

Figure 2: The Innovation Context (Loughnane 2007)

Predictability requires stability. Stability is achieved through policies and procedures coupled with flawless execution and with very few decisions. As chaos is approached, the number of decisions made increases dramati-cally. The complexity of the context increases. Flawless execution becomes more difficult. The power (knowledge, support and resources) required to make good decisions increases. Decisions are subject to chance rather than predictability.

Innovation exists in a context approaching chaos. It is a context where complexity is high and the unpredictable occurs far more frequently than predictable results.

An organization is not innovative because it has ideas, even great ideas. An organization is innovative when it converts the idea into a completely new product or a technological breakthrough that has the potential to change an industry. Joyce et al (2003) tell us that "innovative companies lead the way with industry changing innovations and a willingness to cannibalize offerings, resisting the temptation to wring every last cent out of a product before introducing another to take its place."

3. The Problem

Deciding that innovation is an essential management practice is not a decision to be taken without a great deal of reflection. Few senior leaders have a clear, well-developed model of what innovation looks like as an organizational capability. And since they don't know what it looks like, they don't know how to build it (Hamel, 2003).

Hamel (2003) found two core challenges must be overcome if innovation is to be developed into a deep capability in any organization. The first challenge is that most companies have a very narrow idea of innovation, usually focusing just on products and services. Second, most companies devote much more energy to optimizing what is there (current products, services capabilities) than to imagining what could be.

There is not a shortage of creative people in most modern business organizations. The problem is the shortage of management skills necessary to follow-through. By its very nature the creative process often is not structured nor are the creative people. Many creative people do not know how an organization gets things done. Many creative people do not know how the organization makes money. Theodore Levit (2002) informs us that creativity is not enough:

> "All too often there is a peculiar underlying assumption that creativity automatically leads to actual innovation."

Creativity and innovation that follows often require organizational change. Most organizational-change programs do not achieve their intended results. Why? Consider a simple example. The creative person has lots of ideas but no implementation skills. The next in command person has a high degree of capability to manage the status quo but has a low acceptance of new ideas, a low capability to consider change and a low level of skill to manage change. Perhaps creativity will occur, but not innovation. Few companies with a history of stability can change to innovative companies simply because of the good ideas of creative people and the good intentions of top management.

Organizations, creativity and innovation do not make for a happy marriage. Creativity requires 'permissiveness'. Organizations require order and conformity to get things done. Creativity and innovation can wreak havoc on the organization.

4. The Solution

Suppose that an organization decides that it needs to be creative and innovative. How can the circle created by the problems between the organization and the need to creative and innovative be squared, i.e., what is the solution?

Perhaps one starting point is to define innovation differently. Amar Bhide (2006) discusses two types of innovation: *upstream* as the development of new inventions and technologies; and *downstream* as a system of turning inventions and processes into economic value. Upstream and downstream innovations require different knowledge skills and resources and thus different strategies.

At this point it is important to remember that formulating and implementing strategy is a management practice. Innovation is also a management practice. The ability to practice innovation is the result of the successful implementation of a strategy. Innovation results from innovation process execution. Figure 3 illustrates innovation as a system that lies on a continuum. Innovation is a complex and usually gradual process that involves many players (companies) making incremental advances over time – the continuities. Scientists and engineers work at ideas and creativity at the upstream point. At the downstream point, big ideas are adapted to create economic value at a local level. Along the continuum there are proliferations of species (different firms responding to different customer needs). The species exist in different local competitive environments with different business systems (dimensions).

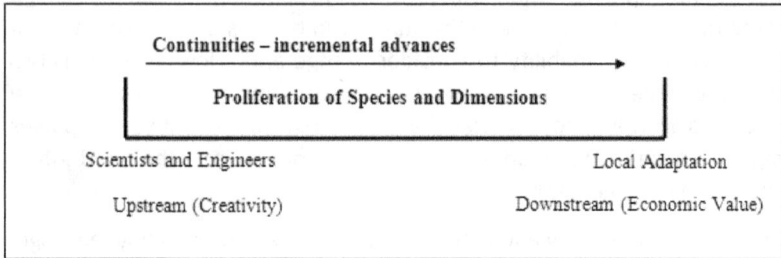

```
Continuities – incremental advances
                                              ⟶
Proliferation of Species and Dimensions

Scientists and Engineers                    Local Adaptation

Upstream (Creativity)              Downstream (Economic Value)
```

Figure 3: The Innovation System (Loughnane 2007)

First, it is important for an organization to decide where it exists along the Innovation System continuum and how it creates value within the system. The organization must determine which secondary management practices – talent, innovation, leadership, mergers and acquisitions - best supports its primary management practices - strategy, execution, culture and structure. Second, it must choose to practice innovation. Third, the organization must assess its competitive environment to determine what level of creativity and innovation is needed as a management practice.

5. Back to the Basics - Strategy, Execution, Culture and Structure – Primary Management Practices

In their groundbreaking study of 200 management techniques Joyce, et al, (2003) found surprising results: most techniques themselves have no direct impact on superior business performance. What does? Mastery of business basics is the key to superior performance.

To sustain superior performance, an organization has to excel at four primary management practices —strategy, execution, culture, and structure—and any two of four secondary practices—talent, leadership, innovation, and mergers and partnerships.

The key to this "4 + 2" formula is not which technique you choose within each practice, but how well and consistently an organization sticks with it. There's no recipe to follow. But the most enduringly successful companies in the study (those delivering a 10-fold return to investors over a 10-year period) clearly demonstrated hallmarks that any organization can follow.

What one learns about strategy from Joyce, et al (2003), is how market competition is shaping up for the future. Innovation is a practice that requires a constant research effort on customer knowledge, market analysis, pricing, the competition, and much more. The authors also reveal their assumptions of how strategy is developed in the organization. Strategy is, they argue, emergent. Henry Mintzberg (1994) would agree that strategies emerge but when looking at results Mintzberg argues that strategies are *realized*. Joyce, et al (2003) persuade us that there is a synthesis between the vision prescribed by the corporation's executives and those actions, investment decisions, and prioritizations that bubble up from the bottom of the organization from among middle managers, engineers, the sales force, and financial staff. Mintzberg (1994) informs us that realized strategies are the result of patterns of decisions over time.

What is important is there is no choice about the primary management practices; however, there is a decision to be made in choosing to pursue innovation as a secondary management practice.

6. Choosing Innovation as a Secondary Management Practices

Vision drives the process of deciding to pursue innovation. Vision is the result of competent strategic analysis (a process). Vision results from an examination of the environment (the context) in which an organization exists. A true vision reflects the direction that is desirable and possible. Choosing to be an innovative organization is a risky decision – it has to be possible.

7. Aligning the Primary and Secondary Management Practices

A continual program to deepen the capability to perform both the primary and secondary management processes is essential. Accomplished athletes make competing look easy, but their skill is not just the result of being born with certain attributes. Their skill comes from practice and discipline over time. Athletic superior performance is the result of a balance of the physical, mental and emotional system. Organizations are systems and the parts of the systems interact. In organizations the alignment of the primary and secondary management practices requires a systems approach. None of the practices are *stand-alone*. The practices must be aligned and accomplished at the same time.

8. The Practice of Innovation

To be successful at innovation an organization must have a "well developed model of what innovation looks like as an organizational capability" (Hamel, 2003). According to Hamel, few senior managers can describe their corporate innovation system. In fact, few companies practice innovation. A review of the literature supports this view. The same companies, Dell, Starbucks, Wal-Mart and Google are examples that are cited as being innovative by multiple authors in multiple articles.

To develop an organizational capability in innovation requires a strategic decision. One must remember at this point innovation is one way an organization will create competitive advantage (value for which a customer will pay). For a strategic decision to be implemented requires a significant investment for which there must be a significant return. Investment in creating innovation capability is about making a trade-off regarding how a company will compete. A company must evaluate trade-offs to determine the greatest return on investment.

(Joyce, et al, 2003) advise that it is not important what technique is used to perform the primary or secondary practices. For example, Sawhney, Wolcott and Arroniz (2006) describe 12 dimensions of the *business system* in which an organization can look for opportunities to innovate. They identify four key dimensions as anchors: offerings (what), customers served (who), processes employed (how) and points of presence it uses to take its offerings to market (where). Hax and Wilde (1999) developed the Delta model. According to Hax and Wilde, the Delta model defines strategic positions that reflect fundamentally new sources of profitability, aligns the strategic positions with a firm's activities, introduces adaptive processes and shows granular metrics that are drivers of performance. What is important is that an organization excels at the technique chosen. Jim Collins reported in *Good to Great* (2001) that what existed in companies that went from good to great was a down-to-earth, pragmatic, commitment to *excellence of process* - a framework - that kept the company, its leaders and its people on track for the long haul.

9. Investing in Innovation

Investing in innovation is a strategic decision. Determining what innovative capabilities a particular company needs is the result of analysis of the company and its environment. Selecting the capabilities to be developed is a strategic decision. Innovation is a management practice requiring company specific knowledge, skills and resources. Innovation is hard, real work that can and should be managed like any other practice.

Two businesses that deal in the same goods and services, in the same territory and with the same clientele, cannot coexist equally (Henderson 1989). Every business is different - a business exists in a unique *context*. As context changes – internal and external – companies must adapt. They must adapt by changing the *content* of innovation. For example, content may shift from product innovation to process innovation. However, the process of innovation (identifying opportunities and applying these ideas to the creation of customer value) is a constant.

10. Conclusion

Even though there is a constant theme in the business press and respected business publications for companies to be innovative, it is not necessary for an organization to be innovative to be highly successful. In fact it is problematic that few companies are capable of excelling at innovation – it is a very difficult management practice. Management practice has *content, context* and *process*. It is contextual simply because what works in one situation (context) can easily fail in another.

There is not a shortage of creative people in most modern business organizations. The problem is the shortage of management skills necessary to follow-through. Companies must invest in the development of management skills to take ideas through to value creation. It is important that a company, first decide where it exists within the Innovation System and how it creates value within the system.

There is no choice about primary management practices; however, there is a decision to be made in choosing to pursue innovation as a secondary management practice. Vision drives the process of deciding to pursue innovation. Innovation is a capability and a company that pursues innovation must have a continual program to deepen the capability to perform both the primary and secondary management practices. To be successful at innovation an organization must have a "well developed model of what innovation looks like as an organizational capability". Investing in innovation is a strategic decision.

Every business is different - a business exists in a unique *context*. Context in a dynamic world is constantly changing. Context changes the requirement for content and process must continually be improved. Because of commoditization and global competition, many companies believe innovation is critical to their future success. These companies must determine what exactly is innovation. Although the subject may be at the top of the agenda, many companies have a mistakenly narrow view of it. Many see innovation as synonymous with new product development or traditional research and development. Companies that do not see that innovation is a capability that must be developed and being innovative as a strategic decision that includes trade-offs do so at their peril.

References

Bhide, A.V. (2006), Venturesome Consumption, Innovation and Globalization, Paper for a Joint Conference of CESifo and the Center on Capitalism and Society, Venice, 21 - 22 July 2006

Collins, Jim (2001) *Good to Great*, HarperCollins: New York

Hamel, Gary (2003), Innovation as a Deep Capability, *Leader to Leader*, No. 27 Winter

Hawawini, Gabriel, Venkat Subramanian and Paul Verdi (2002), Is Performance Driven by Industry factors? A New Look at the Evidence. Wiley InterScience (www.interscience.wiley.com)

Hax Arnoldo C and Dean L. Wilde II (1999), The Delta Model: Adaptive Management for a Changing World, *Sloan Management Review*, Vol. 40, No. 2 pp11-28

Henderson, Bruce (1989), The Origins of Strategy, *Harvard Business Review,* November

Joyce William, Nohria Nitin and Bruce Roberson (2003) *What Really Works: The 4+2 Formula for Sustained Business Success,* Harper Collins, New York

Levit, Theodore (2002), Creativity is Not Enough, *Harvard Business Review* ,August

Mintzberg, Henry (1994) *The Rise and Fall of Strategic Planning*, The Free Press: New York

Mohanbir Sawhney, Robert C. Wolcott and Inigo Arroniz (2006), The 12 Different Ways for Companies to Innovate *Sloan management Review* No. 3, pp. 75-81

Porter, Michael E. (1980), Competitive Strategy: Techniques for Analyzing Industries and Competitors, The Free Press: New York.

Transformational and transactional leadership predictors of the 'stimulant' determinants to creativity in organisational work environments

John D. Politis

Higher Colleges of Technology, Dubai, United Arab Emirates.

First published in the Electronic Journal of Knowledge Management (www.ejkm.com) Vol.2 issue 2, 2004.

Editorial Commentary

It is widely recognized that innovativeness – or the firm's capability to develop new products and processes rapidly and efficiently – is a powerful source of competitive advantage (Wheelwright & Clark, 1994). Organizations need to rethink their management style in order to develop a creative thinking climate, and managers need to adopt a different leadership posture to lead, connect with and inspire knowledge workers.

The chapter by Politis draws on an empirical attempt to determine the predictable leadership style for enhancing creativity and innovation. Transformational leadership is evidenced by inspiring and opening perspectives in which individual creativity develops. It represents a compulsory ingredient to the development of innovation culture and capacity. The empirical analysis emphasizes theo-

45

retical insights and is, on the whole, a solid and relevant contribution to the innovation literature. It underscores the fact that managerial issues, like leader behaviour, are becoming more and more important in shaping the future of innovation.

References

Wheelwright, S. C, & Clark, K. B. (1994). Accelerating the design-build-test cycle for effective product development. International Marketing Review, 11(1), 32-46.

Abstract: This paper examines the relationship between the leadership dimensions associated with Bass's (1985) model, and the 'stimulant' and 'obstacle' determinants of the work environment for creativity. There are three major findings in this research. First, the relationship between transformational and transactional leadership and the 'stimulant' determinants of the work environment for creativity is significant and positive. Second, the 'obstacle' determinants of the work environment for creativity are negatively related with both transactional and transformational leadership. Finally, transformational leadership is more strongly correlated than transactional leadership with the 'stimulant' determinants of the work environment for creativity. Thus, transformational leadership is an increasingly important aspect in today's organisations in creating a corporate culture and the work environment that stimulates employees' creativity and innovation.

Keywords: creative work environment, innovation, knowledge management, organisational creativity, transformational and transactional leadership.

1. Introduction

'Create, innovate or die!' That has increasingly become the rallying cry of today's managers. In a dynamic world of global competition, organisations must innovate and create new products and services and adopt state-of-the-art technology if they are to compete successfully (Kay, 1993; Richards, Foster & Morgan, 1998). In general usage, creativity means the ability of people, and hence the ability of employees, to combine ideas in a unique way or to make unusual associations between ideas (Amabile, 1996; Reiter-Palmon & Illies, 2004). Consequently, organisations need to create a climate that encourages and stimulates employees' creative thinking (Amabile, 1988; Goldsmith, 1996). In other words, organisations must try to remove work and organisational barriers that might impede creativity. By doing so, they may replace employees' traditional vertical thinking with zigzag or lateral thinking and might promote divergent thinking by breaking or even challenging the mental models in an individual, and sometimes treating problems as opportunities (Rickards, 1990).

As a result, researchers have become increasingly interested in studying environmental and work factors conducive to creativity and innovation (Amabile, Conti, Coon, Lazenby & Herron, 1996; Ford, 1996; Oldham & Cummings, 1996). Theory and research suggest that employees have creative potential if we can learn to unleash it. Creative potential might be unleashed when employees are given adequate resources to conduct their work (Delbecq & Mills, 1985), when their work is intellectually challenging (Amabile & Gryskiewicz, 1987), when they are given high level of autonomy and control over their own work (King & West, 1985), and when they given intrinsic task motivation (Robbins, 2003). In relation to leadership and intrinsic motivation, a study by Tyagi (1985) of 168 life insurance salespersons showed that supportive and facilitative leadership accounts for 38 percent of the variance in salespersons' extrinsic motivation and only 16 percent of their intrinsic motivation. Thus, one cannot immovably suggest that supportive leadership will enhance employees' creativity through intrinsic motivation. Moreover, although Amabile and Gryskiewicz (1987) revealed that leader's enthusiasm, interest, and commitment to new ideas and challenges encouraged scientists' creativity, leadership has not been treated as a particularly important influence on creativity (Mumford, Scott, Gaddis & Strange, 2002).

Overall, the literature linking leader behaviours to individual creative performance is scant (Amabile, Schatzel, Moneta & Kramer, 2004), and the literature linking transformational and transactional leadership to work environment dimensions that are most conducive to creativity and innovation is even smaller. To this end, this research started by asking the following questions. To what extent will leaders, who provide adequate resources and delegate authority to their subordinates, affect the determinants of the creative work environment, which in turn, affect creativity and innovation? Which leadership styles best supports the 'stimulant', and which, supports the 'obstacle' determinants of the work environment for creativity. Do leadership behaviours have at all an effect on removing work and organisational barriers that might impede creativity? The answers to these questions are some of the objectives of this paper.

The research reported in this study investigates the relationship between transformational and transactional leadership and the determinants of the creative work environment. The study involves a questionnaire-based survey of members of self-managing teams from a high technology organisation operating in the United Arab Emirates.

2. Literature review

2.1. Models of creativity – the work environment for creativity

Current views on organisational creativity focus on the outcomes or creative products (i.e. goods and services). A creative product is defined as one that is both (a) novel or original and (b) potentially useful or appropriate to the organisation (Amabile, 1996; Ford, 1996; Mumford & Gustafson, 1988). Various factors contribute to the generation of creative products, both at the individual and organisational levels (Mumford & Gustafson, 1988).

At the individual level, an extensive body of research suggests that individual creativity essentially requires expertise, creative-thinking skills, and intrinsic task motivation (Amabile, 1997). Expertise refers to knowledge, proficiencies, and abilities of employees to make creative contributions to their fields. Creative-thinking skills include cognitive styles, cognitive strategies, as well as personality variables that influence the application of these creative-thinking skills. Task motivation refers to the desire to work on something because it is interesting, involving, exciting, satisfying, or personally challenging. Task motivation is crucial in turning creative potential into actual creative ideas (Robbins, 2003). Studies confirm that the higher the level of each of these three components, the higher the creativity.

At the organisational level, researchers have also included individual characteristics as part of the broader framework explaining creativity in the work place. Woodman, Sawyer and Griffin (1993), included personality variables, cognitive factors, intrinsic motivation, and knowledge in their model of organisational creativity. Yet, research in social psychology suggests that supportive behaviour on the part of others in the work place (i.e. co-workers and supervisors) enhances employees' creativity (Amabile et al., 1996; Oldham & Cummings, 1996; Tierney, Farmer and Graen, 1999). Other areas of research have suggested that organisational support and evaluation of new ideas is necessary to encourage employees' creativity (Kanter, 1983). Rewards and bonuses were also reported as essential ingredients in the process of creating a creative work environment (Amabile et al., 1996). Moreover, it has been suggested that there are factors (i.e. internal political problems, conservatism and rigid formal structures) that could impede creativity amongst individuals (Amabile & Gryskiewicz, 1987).

The above literature suggests that individual creativity is a complex phenomenon that is influenced by multiple individual-level variables as well as contextual and environmental variables. The focus then of individual creativity is on the specific contextual variable of leadership and on the theories of organisational creativity – the componential theory of Amabile (1988), the interactionist theory of Woodman et al. (1993), and the multiple social domains theory of Ford (1996) – all of which include the work environment as an influence on employee creativity.

In relation to the environmental variables, Amabile et al.'s (2004) componential theory of creativity is the only theory that specifies *creativity features* that contribute to the perceived work environment for creativity. But, how can organisations assess the dimensions of the perceived work environment that might influence employees' creativity? Amabile and colleagues (1996) have drawn on the literature of creativity and developed an instrument which assesses the dimensions of the work environment that have been suggested in empirical research and theory as essential for organisational creativity. This instrument is referred in the literature as KEYS. Eight determinants (dimensions) of the work environment for creativity are measured by KEYS (Amabile, 1995). Of the eight, six are referred to as 'stimulant' dimensions and have a positive (+) influence on the creative work environment, while the remaining two are referred to as 'obstacle' dimensions and have a negative (-) effect (Amabile et al., 1996). The eight determinants, and the main areas covered by each, are shown in the Appendix.

In relation to leadership it is suggested that leadership is a crucial variable contributing to the culture and climate of the organisation and perception of support for creativity and innovation (Amabile & Gryskiewicz, 1989; Cummings & Oldham, 1997; Mumford, Whetzel & Reiter-Palmon, 1997; Mumford et al. 2002). Therefore, there must be a dynamic interaction between leadership and creativity in a way of supporting, encouraging and energising the perceptions and behaviours of employees that influence the creative work environment.

2.2. Specific leader behaviours and creativity

The literature over the past 30 years has documented the importance of perceived leader support for subordinate creativity (For a review, see Mumford et al., 2002). Studies have demonstrated that team members' collective view of support from their leader is associated with the team's success in creative endeavours (Amabile & Conti, 1999; Amabile et al. 1996). But which leadership style best supports subordinates' creative thinking? Is it the Stogdill's (1974) Ohio Studies of initiating structure and consideration? It is the Blake and Mouton's (1964) task-orientation and relationship-orientation leadership? Is it the Vroom and Yetton's (1973) participative leadership, or the Bass's (1985) transformational and transactional leadership?

A review of the literature suggests that neither the classic Ohio two-factor leadership model, nor the Ekvall (1991) relationship-orientation, and change-orientation leadership, can easily accommodate the facilitator kind of leadership that is needed for creativity. The literature suggests that a leadership role of a facilitative kind fosters the generation of new (creative) outputs (Ekvall, 1991). It is also reported that supportive, no-controlling supervision, enhances creativity (Oldham & Cummings, 1996), and employees are more creative when they are given high levels of autonomy (King & West, 1985). From the above literature one can argue that creative leadership style seems to have much in common with Bass's (1985) transformational leadership (Rickards & Moger, 2000). It is thus, reasonable to expect that the leadership style that focuses on specific techniques, such as, involving employees in the decision-making process and problem-solving, empowering, and supporting them to develop greater autonomy, coaching and teaching them, and helping them to look at old problems in new ways (Burns, 1978; Bass, 1985, 1990), is essential to influence the behaviour of employees in creating a work environment conducive to creativity. The leadership style focusing on such specific techniques is known as 'transformational' leadership. Consequently, the dimensions of transformational and transactional leadership were employed to predict the determinants of the creative work environment.

2.2.1. Transformational and transactional leadership

Transformational and transactional leadership dimensions were derived from Bass's (1985) theory and research. Transformational leaders are those who "inspire followers to transcend their self-interests and who are capable of having a profound and extraordinary effect on followers" (Robbins, 2003: 343). On the other hand, transactional leaders are those who "guide or motivate their followers in the direction of established goals by clarifying role and task requirements (Robbins, 2003: 343). Bass (1985) developed the multifactor leadership questionnaire (MLQ-Form 5), which measures five leadership factors. The five factors tapped by the MLQ-5 include: *charismatic behaviour, individualised consideration* and *intellectual stimulation*, forming the transformational leadership dimension. *Contingent reward* and *management-by-exception (MBE) passive*, forming the transactional leadership dimension. The following definitions are taken from Hater and Bass (1988: 696).

Transformational leadership

- *Charismatic behaviour:* 'the leader instills pride, faith, and respect, has a gift for seeing what is really important, and transmits a sense of mission'.
- *Individualised consideration:* 'the leader delegates projects to stimulate learning experiences, provides coaching and teaching, and treats each follower as individual'.
- *Intellectual stimulation:* 'the leader arouses followers to think in new ways and emphasises problem solving and the use of reasoning before taking action'.
- *Transactional leadership*
- *Contingent reward:* 'the leader provides rewards if followers perform in accordance with contracts or expend the necessary effort'.
- *Management-by-exception passive:* 'the leader avoids giving directions if the old ways are working and allows followers to continue doing their jobs as always if performance goals are met'.

A review of the literature suggests that subordinates' creativity is a function of their perceptions of the general work environment for creativity, which is, in turn, a function of their relationship with the leader; a leader who is characterised by trust, mutual linking, and respect (Zhou & Shalley, in press). The foundation of creative leadership then is based on specific leader behaviours akin to relationship-oriented ("consideration") and transformational leadership (Rickards & Moger, 2000). Moreover, Jones

(1996) suggested that the leader with hierarchical attitudes (i.e. diametrically opposite to creative leader) will create a rigid formal structure which blocks dialogue and hence creativity. It is thus reasonable to hypothesise that the factors representing the 'stimulant' components of the creative work environment will be more strongly, and more positively correlated with the factors of transformational leadership, than will be the factors representing the 'obstacle' components of the creative work environment. The assumed connectedness between transformational leadership and the determinants of the work environment for creativity is expressed in Hypothesis 1.

Hypothesis 1: Correlations between each of the transformational leadership behaviours and the 'stimulant' determinants of the creative work environment will be stronger, and more positive, than those with the 'obstacle' determinants of the creative work environment.

Moreover, Amabile and colleagues (2004) have provided empirical evidence suggesting that team leader supportive behaviour, which includes both task-oriented and relationship-oriented support, is an important aspect of the perceived work environment for creativity. It is thus plausible to predict that the factors representing the 'stimulant' components of the creative work environment will be more strongly, and more positively correlated with the factors of transactional leadership, than will be the factors representing the 'obstacle' components of the creative work environment. The assumed connectedness between transactional leadership and the determinants of the work environment for creativity is expressed in Hypothesis 2.

Hypothesis 2: Correlations between each of the transactional leadership behaviours and the 'stimulant' determinants of the creative work environment will be stronger, and more positive, than those with the 'obstacle' determinants of the creative work environment.

3. Subjects and procedure

3.1. Sample and procedures

Sample: The study focused in a service organisation operating in the United Arab Emirates (UAE). Nine departments involved in communications technology have participated in the study, all of which are recognised for their creativity. Respondents were full-time employees of the participating departments and volunteered to participate in the study. Question-naires, written in English, containing items measuring the determinants of the creative work environment and the dimensions of transforma-tional/transactional leadership were distributed to 173 members of self-managing teams in the nine departments. One hundred eighteen (118) employees returned usable questionnaires; yielding a 68 percent response rate. Most were from the new product development (57 percent), and cus-tomer service (17 percent) departments. The remaining ones were spread among various other areas including education/training, consulting, etc (26 percent). The majority were within the 21-30 age group (81 percent). Given the relatively young age of the sample, the level of work experience is accordingly low. Eighty two (82) percent of the respondents have had five or less years of work experience. The respondents were 6 percent fe-male and 94 percent males and all had either a technical or university qualification taught in the English language. Anonymity was guaranteed and no names or other identifying information was asked.

Procedures: Survey questionnaires were pre-tested, using a small number of respondents (about one dozen; the pre-test participants did not partici-pate in the final data collection). As a consequence of the pre-testing, rela-tively minor modifications were made in the written instructions and in several of the demographic items. The revised survey was then adminis-tered to the respondents of the nine departments in their natural work settings. Written instructions, along with brief oral presentations, were given to assure the respondents of anonymity protection and to explain (in broad terms) the purpose of the research. The participants were all given the opportunity to ask questions and were encouraged to answer the sur-vey honestly; anonymity was guaranteed and no names or other identify-ing information was asked.

3.1.1. Analytical procedure

Confirmatory factor analyses (CFAs) were performed using the analysis of moment structures (AMOS, version 5) software (Arbuckle, 2003) for the factor analysis of the measurement models. Using CFAs, we assessed the validity of the measurement models of the variables used in the paper. A mixture of fit-indices was employed to assess the overall fit of the meas-urement models. The ratio of chi-square to degrees of freedom (χ^2/df) has been computed, with ratios of less than 2.0 indicating a good fit. However, since absolute indices can be adversely effected by sample size (Loehlin, 1992), four other relative indices, the goodness-of-fit index (GFI), the adjusted goodness-of-fit index (AGFI), the comparative fit index (CFI), and the Tucker and Lewis index (TLI) were computed to provide a more robust evaluation of model fit (Tanaka, 1987; Tucker & Lewis, 1973). For GFI, AGFI, CFI and TLI, coefficients closer to unity indicate a good fit, with acceptable levels of fit being above 0.90 (Marsh, Balla & McDonald, 1988). For root mean square residual (RMR), and root mean square error approximation (RMSEA), evidence of good fit is considered to be values less than 0.05; values from 0.05 to 0.10 are indicative of moderate fit and values greater than 0.10 are taken to be evidence of a poorly fitting model (Browne & Cudeck, 1993).

Given adequate validity of those measures, we reduced the number of indicator variables by creating a composite scale for each latent variable (Politis, 2001). These scales were subjected to a series of correlational and regression analysis.

4. Results

4.1. Measurement models

The variables that we measure on the survey are: transformational and transactional leadership, and the determinants of the work environment for creativity.

4.1.1. Independent variables

Transformational and *transactional* leadership measures were assessed using Bass's (1985) 73-item multifactor leadership questionnaire (MLQ–Form 5). The MLQ-5 questionnaire employs a 5-point response scale (0 = not at all; 4 = frequently if not always) and consists of five subscales: three subscales forming the transformational leadership (i.e. charismatic behaviour, individualised consideration, and intellectual stimulation), and two subscales forming the transactional leadership (i.e. contingent reward and management-by-exception). We conducted CFA of all MLQ items in order to check for construct independence .We first fit a five-factor model to the data, corresponding to that proposed by Bass. The fit indices of CFI, AGFI, CFI, TLI, RMR, and RMSEA were 0.91, 0.96, 0.97, 0.89, 0.05, and 0.07, respectively, suggesting that the five factor model provides a good fit. Thus, the data supported the independence of five factors, namely, charismatic behaviour (α = 0.91); individualised consideration (α = 0.85); intellectual stimulation (α = 0.78); contingent reward (α = 0.87); and management-by-exception (α = 0.67). Twelve items of the MLQ were dropped due to cross loading and/or poor loading of the order of, or less than 0.11.

4.1.2. Dependent variables

Determinants of the work environment for creativity made up of eight sub-categories, namely, organisational encouragement, supervisory encouragement, work group supports, freedom, sufficient resources, challenging work, workload pressure, and organisational impediments. These categories were assessed using Amabile et al.'s (1996) 66-item instrument (KEYS). The instrument employs a 4-point response scale (1 = never; 4 = always). We conducted CFA of all KEYS items in order to check for construct independence. We first fit an eight-factor model to the data, corresponding to that proposed by Amabile et al. (1996). The fit indices of CFI, AGFI, CFI, TLI, RMR, and RMSEA were 0.88, 0.90, 0.93, 0.89, 0.06, and 0.08, respectively, suggesting that the eight factor model provides a reasonable fit. Thus, the data supported the independence of eight factors, namely, organisational encouragement (8 items, α = 0.83), supervisory encouragement (7 items, α = 0.85), work group support (8 items, α = 0.77), freedom (3 items, α = 0.67), sufficient resources (5 items, α = 0.72), challenging work (4 items, α = 0.81), workload pressure (3 items, α = 0.80), and organisational impediments (7 items, α = 0.72). Twenty one items of the KEYS were dropped due to cross loading and/or poor loading of the order of, or less than 0.08.

Moreover, for the purpose of this study we created a "stimulant" index to creativity by averaging scores for organisational encouragement, supervisory encouragement, work group support, freedom, sufficient resources, and challenging work items (α = 0.88). In addition, we averaged scores from workload pressure and organisational impediments items to form the "obstacle" index to creativity (α = 0.71). The model of Figure 1 summarises the variables used in this paper.

Transformational/transactional leadership dimensions	Dimensions of the creativity work environment
Transformational & Transactional Leadership (Bass, 1985) Transformational Leadership • Charismatic behaviour • Individualised consideration • Intellectual stimulation Transactional Leadership • Contingent reward • Management-by-exception	**Determinants of the work environment for creativity** (Amabile et al., 1996) • Stimulant factors (+) • Organisational encouragement • Supervisory encouragement • Work group support • Freedom • Sufficient resources • Challenging work • Obstacle factors (-) • Workload pressure • Organisational impediment

Figure 1: Summary of variables used in the paper

4.2. Hypothesis testing

Correlation analysis was used to examine the patterns of relationship between the leadership style dimensions and the determinants of the work environment for creativity. Table 1 reports the means, standard deviations, and the correlations among all variables included in the analyses.

There are several important observations regarding Table 1. First, it can be noted that all sub-scales display acceptable reliabilities, these being of the order of, or above, the generally accepted value of 0.70 (Hair, Anderson, Tathan & Black, 1995), with the exception of management-by-exception (α = 0.67). Second, the correlations between the constructs used in this study are generally lower than their reliability estimate, indicating good discriminant validity for these factors (Hair, et al., 1995).

[a] N = 118 individuals of self managing teams; [b] Coefficient alphas (αs) are located along the diagonal.

All correlations above 0.17 are statistically significant, $p < 0.01$; all correlations between 0.15 and 0.16 are statistically significant, $p < 0.05$.

As shown in Table 1, both hypotheses are supported by this data for both dimensions of the work environment for creativity. As predicted, the three transformational leadership variables showed significant correlations with the *stimulant* factors of creativity. The results indicate that the correlations between transformational leadership variables and the stimulant determinants of creativity are stronger, and more positive, than those with the obstacle determinants of creativity, supporting Hypothesis 1. (In fact, the correlations with the obstacle determinants of creativity are negative and non-significant.) Specifically, the results showed strong positive correlations between the stimulant factors of creativity and charismatic behaviour ($r = 0.26$, $p < 0.01$); individualised stimulation ($r = 0.38$, $p < 0.01$); and intellectual stimulation ($r = 0.31$, $p < 0.01$). Moreover, the results showed non-significant and negative correlations between the obstacle determinants of creativity and charismatic behaviour ($r = -0.16$); individualised stimulation ($r = -0.09$); and intellectual stimulation ($r = -0.15$).

Table 1: Means, standard deviations, and correlations of leadership and the determinants of the work environment for creativity

Latent variable	Mean[a]	SD	1	2	3	4	5	6	7
Transformational leadership									
1. Charismatic behaviour	1.93	1.08	.91[b]						
2. Individualised consideration	2.07	1.03	.82	.85					
3. Intellectual stimulation	2.01	1.06	.76	.69	.78				
Transactional leadership									
4. Contingent reward	1.91	1.05	.80	.84	.75	.87			
5. Management-by-exception (passive)	2.19	0.72	-.20	-.25	-.09	-.16	.67		
Determinants of the creative work environment									
6. Stimulant determinants for creativity	2.71	0.49	.26	.38	.31	.22	.15	.88	
7. Obstacle determinants for creativity	2.71	0.57	-.16	-.09	-.15	-.09	-.04	-.26	.71

Furthermore, results indicate that the correlations between transactional leadership variables and the stimulant determinants of creativity are stronger, and more positive, than those with the obstacle determinants of creativity, supporting Hypothesis 2. The results showed moderate positive correlations between the stimulant factors of creativity and contingent rewards ($r = 0.22$, $p < 0.01$); and management-by-exception ($r = 0.15$, $p < 0.05$), and negative, near zero, and non-significant correlations between the obstacle determinants of creativity and contingent rewards ($r = -0.09$); and management-by-exception ($r = -0.04$).

In view of significant correlations between the variables, further tests were performed to identify the main factors affecting the determinants of the creative work environment. This analysis was performed using regression models. The regression results indicated that the transformational variables jointly (i.e. charismatic behaviour, individualised stimulation, and intellectual stimulation) explained nearly a third variance of the stimulant factors of creativity (R-square = 0.29, F = 4.7, $p < 0.01$), while the transactional variables alone (i.e. contingent rewards, and management-by-exception) explained only 9% of the variance (R-square = 0.09, F = 7.1, $p < 0.05$). (Note that both of the independent variables jointly (i.e. transformational and transactional) explained just over a third variance of the stimulant factors of creativity (R-square = 0.34, F = 3.6, $p < 0.01$.)) There was no significant direct effect found of the transformational and transactional variables towards the obstacle factors of creativity (R-square = 0.07, F = 2.16, $p > 0.05$; R-square = 0.02, F = 1.17, $p > 0.05$, respectively).

5. Discussion

The need of organisations to be more competitive has sparked the interest of researches and practitioners to understand creativity in the workplace (Mumford et al., 2002). This study examined specific contextual variables of leadership and environmental variables that are conducive to creativity and innovation. Although replication of all research results is certainly desirable, the current study seems to highlight that both transformational and transactional leadership behaviour impact of the stimulant (i.e. organisational encouragement, supervisory encouragement, work group support, freedom, sufficient resources, and challenging work) determinants of the work environment conducive to creativity in an organisation (communications technology) which is recognised for its creativity. The findings are consistent with the realm of supportive management style and employees' creative performance theories. The results of the study reinforce the componential theory of Amabile (1988), and indeed go beyond prior research of particular areas of leader support, such as the leader's tendency to provide both clear strategic direction and procedural autonomy in carrying out the work (Pelz & Andrews 1976), or supportive, no-controlling supervision (Oldham & Cummings, 1996).

Transformational and transactional leadership predictors of the 'stimulant' determinants to creativity in organisational work environments

The key finding of this study is undoubtedly that the leaders, who see what is important, transmit a sense of mission, provide coaching/teaching, and arouse employees to think in new ways and emphasise problem solving, are most effective in facilitating the stimulant determinants of the creative work environment, as established by Amabile et al. (1996). Specifically, the three transformational leadership variables alone explained over 29% of the variance of the stimulant determinants of creativity. This finding is particularly significant and important in the work environment for creativity landscape that is rich in theory and rhetoric, but scarce in empirical evidence. The findings suggest that it is those particular transformational leader behaviours (i.e. charismatic behaviour, individualised consideration and intellectual stimulation) that appear to have the impact on the perceived work environment that influence employees' creative freedom, encouragement and intrinsic motivation for creativity. These leadership behaviours are indeed essential in the process of creating new knowledge, applying knowledge and in the words of Peter Druker (1993) "making it productive".

Furthermore, it is also important to note that the remaining 71% of the variance is not explained by the variables tested in this study. One could assume that a portion of the remaining variance could be explained by other leadership styles, such as Stogdill's (1974) consideration leadership, and Manz and Sims's (1987) self-management leadership, both of which contain certain themes common to those measured by Bass's (1985) transformational leadership dimensions. In addition, another portion of the remaining variance could be explained by the subordinates' perceptions of themselves – particularly their competence and the value of their work (Amabile et al., 2004), the employees' mood (Isen, 1999); and the employees' personality characteristics (Amabile, 1996; Feist, 1999). Thus, future research should examine models that integrate the Ohio studies consideration leadership; the self-management leadership factor of the Manz and Sims's (1987) studies; the transformational/transactional leadership factors of the Bass's (1985) studies; the variables of personality characteristics; employee's mood; and the subordinates' perceptions of themselves.

This study also has implications for theories of leader behaviour. The classic two-factor theory of leader bahaviour (Fleishman, 1953) proposes that effective leaders must engage in both task and relationship management (i.e. initiating structure and consideration behaviours). Our findings showed that transformational leadership (comparable to consideration behaviour) is a better predictor of the stimulant determinants of the creative work environment than transactional leadership (comparable to initiating structure). It appears that *effective* creative leadership requires skills not only in managing both subordinate tasks and subordinates relationship, but also in integrating the two simultaneously. Moreover, our findings indeed support the superiority of transformational over transactional leadership behaviour (Politis, 2002).

In summary, the results of this study have shown that (a) there is a positive and significant relationship between transformational/transactional leadership and the stimulant determinants of the work environment for creativity; (b) the factors representing transformational leadership are better predictors of the stimulant determinants of the creative work environment than those of transactional leadership; and (c) the obstacle determinants of the work environment for creativity are negatively associated with both transformational and transactional leadership.

6. Limitations and future work

While this research has established a clear relationship between transformational and transactional leadership and the stimulant factors to creativity, some caution must be exercised when interpreting these findings due to a number of limiting factors. First, although a quantitative study is able to establish a relatively clear picture of relationships between phenomena, it is less apt at explaining the reasons behind it. Thus, future qualitative research needs to be considered to explore the exact reasons why transformational/transactional leadership tends to lead to stronger associations with the stimulant determinants of the work environment for creativity than with the obstacle determinants for creativity. Other limitations include the use of a relatively undeveloped instrument measuring the perceptions of the creative work environment (note: 21 items were dropped from the KEYS measurement model due to cross or poor loading), inability to establish causality, and the relatively small sample size.

References

Arbuckle, J. L. (2003) *Analysis of moment structures (AMOS), user's guide version 5.0,* SmallWaters Corporation, Chicago, IL.

Amabile, T. M. (1997) "Motivating creativity in organisations: On doing what you love and loving what you do", *California Management Review,* Vol. 40, pp39-58.

Amabile, T. M. (1996) *Creativity in context,* Westview Press, Boulder, CO.

Amabile, T. M. (1995), KEYS User's Manual: Assessing the climate for creativity, Centre for Creative Leadership, PO Box 16300, Greensboro, North Carolina, 27438-6300, USA.

Amabile, T. M. (1988) "A model of creativity and innovation in organisations", in *Research in Organisational Behaviour,* B. M. Staw and L. L. Cummings (Eds), 10 CT: JAI Press, Greenwich, pp123-167.

Amabile, T. M. & Conti, R. (1999) "Changes in the work environment for creativity during downsizing", *Academy of Management Journal,* Vol 42, pp630-640.

Amabile, T. M. Conti, R. Coon, H. Lazenby, J. & Herron, M. (1996) "Assessing the work environment for creativity", *Academy of Management Journal,* Vol 39, pp1154-1184.

Amabile, T. M. & Gryskiewicz, S. S. (1987) *"Creativity in the R &D laboratory",* Technical Report No. 30, Center for Creative Leadership, Greensboro, NC.

Amabile, T. M. & Gryskiewicz, N. D. (1989) "The creative environment scales: Work environment inventory", *Creativity Research Journals,* Vol 2, pp231-254.

Amabile, T. M. Schatzel, E. A. Moneta, G. B. & Kramer, S. J. (2004) "Leader behaviours and the work environment for creativity: Perceived leader support", *The Leadership Quarterly,* Vol 14, pp5-32.

Bass, B. M. (1985) Leadership and performance beyond expectations, Free Press, NY.

Bass, B. M. (1990) Bass and Stogdill's handbook of leadership: Theory, research, and managerial applications, Free Press, NY.

Blake, R. R. & Mouton, J. S. (1964) *The managerial grid,* Gulf Publishing Company, Houston, TX.

Browne, M. W. & Cudeck. R. (1993) "Alternative ways of assessing model fit" in *Testing Structural Equations Models,* Bollen, K. A. and Scott Long, J. (Eds), Sage, Newbury Park, California, pp36–62.

Burns, J. M. (1978) *Leadership,* Harper & Row, NY.

Cummings, A. & Oldham, G. R. (1997) "Enhancing creativity: Managing work contexts for the high potential employee", *California Management Review,* Vol 40, pp22-39.

Delbecq, A. L. & Mills, P. K. (1985) "Managerial practices and enhance innovation", *Organisational Dynamics,* Vol 14, No.1, pp24-34.

Druker, P. F. (1993) *Post-capitalistic society,* Butterworth-Heinemann, Oxford.

Ekvall, G. (1991) "The organisational culture of idea management: A creative climate for the management of ideas" in *Managing Innovation,* J. Henry and D. Walker (Eds), Sage Publications, London, pp73-79.

Feist, G. J. (1999) "The influence of personality on artistic and scientific creativity" in *Handbook of Creativity,* R. Sternberg (Ed), Cambridge, Cambridge University Press, UK, pp273-296.

Fleishman, E. A. (1953) "The description of supervisory behaviour", *Journal of Applied Psychology,* Vol 37, No.1, pp1-6.

Ford, C. M. (1996) "A theory of individual creative action in multiple social domains", *Academy of Management Review,* Vol 21, pp1112-1142.

Goldsmith, C. (1996) "Overcoming roadblocks to innovation", *Marketing News,* Vol 30, No.24, p 4.

Hair, J. F., Anderson, R. E., Tathan, R. L. & Black, W. C. *Multivariate data analysis with readings,* (4th Edition). Prentice Hall, Englewood Cliffs, New Jersey, 1995.

Hater, J. J. & Bass, B. M. (1988) "Superior's evaluations and subordinate's perceptions of transformational and transactional leadership", *Journal of Applied Psychology,* Vol 73, No.4, pp695–702.

Isen, A. M. (1999) "On the relationship between affect and creative problem solving" in *Affect, Creative Experience and Psychological Adjustment,* S. Russ (Eds), Brunner/Mazel, Philadelphia, pp3-17.

Jones, S. (1996) *Developing a learning culture,* McGraw-Hill, London.

Kanter, R. M. (1983) *The change masters,* Simon and Schuster, NY.

Kay, J. (1993) *Foundations of corporate success,* Oxford University Press, NY.

King, N. & West, M. A. (1985) *Experiences of innovation at work,* SAPU Memo No. 772, University of Sheffield, England.

Loehlin, J. (1992) *Latent variables models,* Erlbaum, Hillside, NJ.

Manz, C. C. & Sims, H. P. Jr. (1987) "Leading workers to lead themselves. The external
leadership of self-managing work teams", *Administrative Science Quarterly*, Vol 32, pp106-129.

Marsh, H. W. Balla, J. R. & McDonald, R. P. (1988) "Goodness-of-fit indexes in
confirmatory factor analysis: The effect of sample size", *Psychological Bulletin*, Vol 103, No.3, pp391-410.

Mumford, M. D. & Gustafson, S. B. (1988) "Creativity syndrome: Integration, application, and innovation", *Psychological Bulleting,* Vol. 103, pp27-43.

Mumford, M. D. Scott, G. M. Gaddis, B. & Strange, J. M. (2002) "Leading creative people: Orchestrating expertise and relationships", *The Leadership Quarterly,* Vol 13, pp705-750.

Mumford, M. D. Whetzel, D. L. & Reiter-Palmon, R. (1997) "Thinking creatively at work: Organisational influence on creative problem solving", *Journal of Creative Behaviour,* Vol 31, pp7-17.

Transformational and transactional leadership predictors of the 'stimulant' determinants to creativity in organisational work environments

Oldham, G. R. & Cummings, A. (1996) "Employee creativity: Personal and contextual factors at work", *Academy of Management Journal,* Vol 39, pp607-634.

Pelz, D. C. & Andrews, F. M. (1976) *Scientists in organisations: Productive climates for research and development,* Institute for Social Research, Ann Arbor, MI.

Politis, J. D. (2001) "The relationship of various leadership styles to knowledge management", *The Leadership and Organizational Development Journal,* Vol 22, No.8, pp354-364.

Politis, J. D. (2002) "Transformational and transactional leadership enabling (disabling) knowledge acquisition of self-managed teams: the consequences for performance", *The Leadership and Organizational Development Journal,* Vol 23, No.4, pp186-197.

Reiter-Palmon, R. & Illies, J. J. (2004) "Leadership and creativity: Understanding leadership from the creative problem-solving perspective", *The Leadership Quarterly,* Vol 15, pp55-77.

Richards, I. Foster, D. & Morgan, R. (1998) "Brand knowledge management: Growing brand equity", *Journal of Knowledge Management,* Vol 2, No.1, pp47-54.

Rickards, T. (1990) Creativity and problem solving at work, Gower, Aldershot.

Rickards, T. & Moger, S. (2000) "Creative leadership processes in project team development: An alternative to Tuckman's stage model", *British Journal of Management,* Vol 11, pp273-283.

Robbins, S. P. (2003) *Organisational behaviour,* 10th ed., Prentice Hall, Inc.

Stogdill, R. M. (1974) Handbook of leadership: A survey of the literature, Free Press, NY.

Tanaka, J. S. (1987) "How big is enough? Sample size and goodness-of fit in structural equations models with latent variables", *Child Development,* Vol 58, pp134-146.

Tierney, P. Farmer, S. M. & Graen G. B. (1999) "An examination of leadership and employee creativity: The relevance of traits and relationships", *Personnel Psychology,* Vol 52, pp591-620.

Tucker, L. R. & Lewis, C. (1973) "The reliability coefficient for maximum likelihood factor analysis", *Psychometrika,* Vol 38, pp1-10.

Tyagi, P. K. (1985) "Relative importance of key job dimensions and leadership behaviours in motivating salesperson work performance", *Journal of Marketing,* Vol 49, pp76-86.

Vroom, V. H. & Yetton, P. W. (1973) *Leadership and decision making,* University of Pittsburgh, Press, Pittsburgh.

Woodman, R. W. Sawyer, J. E. & Griffin, R. W. (1993) "Toward a theory of organisational creativity", *Academy of Management Review,* Vol 18, pp293-321.

Zhou, J. & Shalley, C. E. (in press) Research on employee creativity: A critical review and directions for future research, *Research in Personnel and Human Resources Management.*

63

Appendix

Main areas of each determinant of the creative work environment

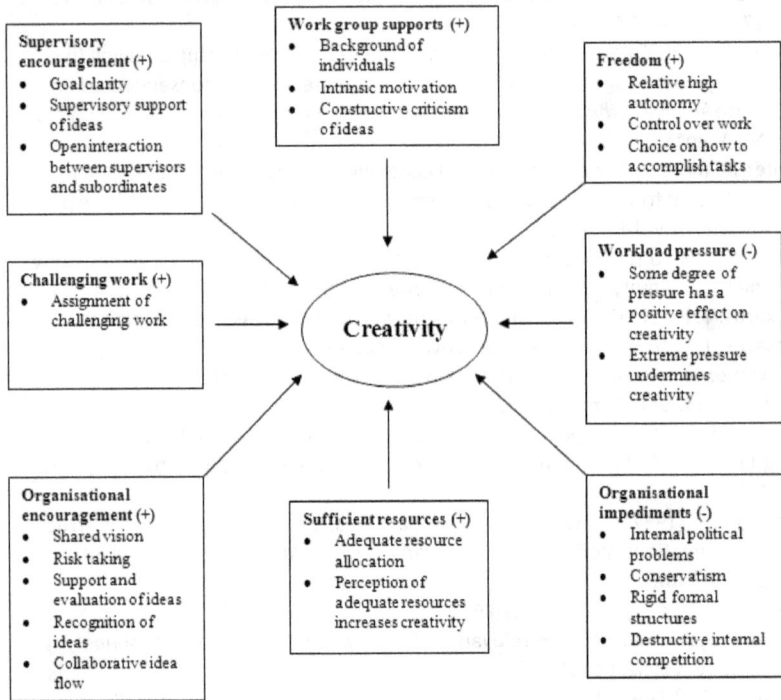

Supervisory encouragement (+)
- Goal clarity
- Supervisory support of ideas
- Open interaction between supervisors and subordinates

Work group supports (+)
- Background of individuals
- Intrinsic motivation
- Constructive criticism of ideas

Freedom (+)
- Relative high autonomy
- Control over work
- Choice on how to accomplish tasks

Challenging work (+)
- Assignment of challenging work

Creativity

Workload pressure (-)
- Some degree of pressure has a positive effect on creativity
- Extreme pressure undermines creativity

Organisational encouragement (+)
- Shared vision
- Risk taking
- Support and evaluation of ideas
- Recognition of ideas
- Collaborative idea flow

Sufficient resources (+)
- Adequate resource allocation
- Perception of adequate resources increases creativity

Organisational impediments (-)
- Internal political problems
- Conservatism
- Rigid formal structures
- Destructive internal competition

Adopted from Amabile et al. (1996)

Note:

- 'Stimulant' determinants of the creative work environment denoted with (+).
- 'Obstacle' determinants of the creative work environment denoted with (-).

Business Benefits of Non-Managed Knowledge

Sinead Devane[1] and Julian Wilson[2]

[1]Bournemouth University, UK
[2]James Wilson (Engravers) Ltd, Poole, UK

First published in the Electronic Journal of Knowledge Management (www.ejkm.com) Vol 7 issue 1, 2009.

Editorial Commentary

Knowledge Management is recognized for its ability to improve organizational effectiveness and thus, value creation. Basadur & Gelate (2006: 52) have developed a framework which suggests that the process of innovative thinking is correlated with two dimensions, the apprehension of knowledge and its creative utilization through a dynamic tension between the polar opposites on each dimension. But the difficulty resides in the way knowledge is handled since, as it noted in this chapter, it is difficult to manage: "...knowledge exists between two ears, and only between two ears." (Drucker, cited by Kontzer, 2001).

Devane and Wilson present an interesting set of questions on the value of tacit Knowledge Management by arguing that by nature, knowledge is an inextricable part of the individual and cannot be managed as an object. They introduce the notion of agency as a useful concept in acknowledging the potential of individuals to unleash their knowledge. The chapter advocates the management of intrinsic knowledge through assessment of outcomes and consequently the managerial necessity of efficient individual skill development. This leads to a different approach to KM that enhances employees' responsibility and fosters unmanaged knowledge in support of innovativeness.

References

Basadur M. and Gelade G, 2006, The Role of Knowledge Management in the Innovation Process, Creativity And Innovation Management, Volume 15 No. 1.

Abstract: This paper proposes that knowledge cannot be effectively managed, rather that people are so complex and the knowledge they acquire so varied and, even in a room where everyone is being taught the same knowledge, uptake is so disparate that to truly manage what an individual knows is impossible. An effective alternative which allows people to maximise their knowledge (even the bits we might not know they know) is to measure the outputs of an individual: what they achieve. The case study in this paper illustrates one innovative company's design to maximise the knowledge of their employees and how the management-less structure they recently adopted has had a profound effect on the engagement of their workforce - in their work and on their profitability.

The argument will critique the theory of knowledge transfer as the movement of a body of knowledge from one place to another. Additionally it confronts the misuse of the work of Polanyi where theorists have crudely bunched together notions of understanding and the hidden aspects of our knowledge under the title 'tacit' knowledge and juxtaposed these with 'explicit' knowledge, which refers to documented and shared knowledge within an organisation. This paper promotes an approach that follows Coverdale's (psychologist) idea that getting things done by developing skills as opposed to focusing on traditional knowledge management is the premise of the company's success. The anthropological notion of cultural memes is explored, and the idea of agency will be introduced as a useful concept in acknowledging the potential of individuals to unleash their knowledge.

The case study presented here is from empirical ethnographic research within a company in the South West. The company has transitioned from a traditionally managed SME with a hierarchical structure, to a de-centred model for workplace where each individual within the company is responsible for what is essentially like their own mini company within the larger one. The company data (quantitative) from before these changes were introduced compared to the data now (on aspects such as Quality, Delivery and Profitability) tell a remarkable story about the effects of management and organisational structure on individual performance and commercial profitability. It would seem that the best results commercially and in individual competency so far point to a counter-intuitive, hands-off approach to KM within an organisation. They point to what is in fact the non-management of knowledge in the sense that knowledge will not be prescribed or transferred from one person to another, but rather drawn out of the individual.

Keywords: knowledge management, outcomes and application, reification, cultural memes, agency, innovation

1. Introduction

This paper will be presented in three sections; following the introduction, the second section is a theoretical discussion surrounding the nature of knowledge and its management, while the third section will detail our observations in the workplace as a case study, the final concluding section will recommend an approach to knowledge management, or non-management, for organisations that is extremely cost effective, counter-intuitive as it may seem.

Our arguments come with two basic assumptions. The first is our understanding of knowledge: knowledge to us is something infinite and intangible, something unique to each individual. Knowledge itself cannot be measured at a specific time or place. It is only recognised by its implications; that is we only know of its existence through the outcomes it produces.

The second is our use of the term 'conventional management'. By this we refer to top-down hierarchically organised systems, with separate specialised functions for different aspects of the organisation's needs that are prevalent in our business world. We acknowledge there is a diversity of management systems, even within the top-down model, but for the purposes of this paper 'conventional management' refers to a system whereby some individuals have positions of authority over others, and responsibility resides primarily with those in the positions of authority.

2. Theories of knowledge management

Let us first define our understanding of the meaning of the terms 'knowledge', 'management' and 'knowledge management'. Theories of knowledge are extensive, but two of the most quoted and influential theorists are Michael Polanyi and Ikujiro Nonaka. Polanyi separates knowledge into two distinct types: the explicit knowledge as codified and written down, that can be passed on from one person to another; and the tacit knowledge that an individual draws on subconsciously or unconsciously from a range of conceptual and sensory information and images that can be brought to bear in an attempt to make sense of something' (Smith 2003).

For Polanyi, 'tacit knowing achieves comprehension by indwelling, and ... all knowledge consists of or is rooted in such acts of comprehension' (Polanyi, 1958).

Nonaka (1991) takes this position further to say that tacit knowledge can be made explicit (as expressed by Wilson 2002) and corporations who tap into tacit knowledge of the individuals working for them are more success- ful. Within such literature (on systems of management that can capture the knowledge of individuals and share it throughout the organisation), many contributors have lost sight of the value of tacit knowledge and the very fact that it cannot be codified.

Management is usefully defined by Henri Foyal (1967) as comprising of five functions: planning, organising, leading, co-ordinating and controlling. But management can be more effectively defined to avoid the assumption that it must be directed from an individual/group in a position of authority unto an individuals or groups in lower positions. Mary Parker Follett (1868-1933), who wrote on the topic of management in the early twentieth cen- tury, defined management as 'the art of getting things done through peo- ple'. This definition does not imply didactic management from persons in authoritarian positions, but allows for the self-organisation of individuals or groups.

One definition of knowledge management (KM) from the corporate per- spective is that 'KM is the practice of harnessing and exploiting intellectual capital to gain competitive advantage and customer commitment through efficiency, innovation and faster more effective decision-making' (Barth 2000).

The International Journal of Management Science, OMEGA, recently ran a special edition on knowledge management. In a review of this edition, King et al. (2006) outlined a broad perspective to take when examining the dif- ferent contributions to the issue: that knowledge management can be de- scribed in terms of a life cycle. Knowledge is generated in different ways, and stored within organisations, who then transfer or 'share' this knowl- edge with others (employees), where it is used or applied in problem solv- ing. King et al. claim that all contributions to the debate fall somewhere along the cycle.

Our paper critiques the notion that knowledge can be collected and stored in the organisation, extrinsic to the individuals who form the organisation. And then disseminated whole and complete to others.

Some studies (Kjaergaard and Kautz 2008; Orzano et al. 2008) show that, even with the best intentions and optimum conditions, KM implementation programs simply do not work. Kjaergaard and Kautz concluded that the failure of a KM implementation programme lay in the ideological misfit that employees experienced between what they imagined working with KM in their company would be like and the reality of working with KM in the company, thus the process of KM installation was abandoned. Orzano et al. showed how expensive formal technological KM sharing practices that were installed within a medical practice were not efficient (as promised), and in fact lost knowledge in some instances. They juxtaposed these observations with instances where more traditional face to face knowledge sharing practices which were proving to be worth-while.

All of the above treats knowledge as if it is something real and tangible like an asset. Are we guilty of the reification of knowledge?

To reify is to regard or treat (an abstraction) as if it had a concrete or material existence (Wikipedia.com). When knowledge is reified for the purpose of corporate KM, it becomes fixed and tangible. This does not allow for the individual's internal landscape into which any new knowledge must be accommodated. Either the knowledge must be modified to fit an individual's internal landscape, or the individual's internal landscape must be modified to fit the knowledge, or the knowledge is lost (through lack of understanding). Then the process of recall and practice of this knowledge is itself subject to problems during reconstruction. If the knowledge is no longer valid, or circumstances do not support it, then it quickly will be extinguished. The transfer of knowledge is not reliable, repeatable and dependable. The "knowledge" presented to individuals must be interpreted (or not), assimilated (or not), stored (or not), recalled (or not) and applied (or not) in achievement of a goal (or not). This provides ample opportunities for knowledge management to go awry, even given the best intentions of all concerned.

If knowledge is an abstraction, truly knowing what knowledge someone possesses is impossible. And thus it is impossible to manage what someone knows.

As one keynote speaker for the Delphi Group's Collaborative Commerce Summit once scoffed: *'You can't manage knowledge... Knowledge is between two ears, and only between two ears.'* (Drucker, cited by Kontzer 2001). And Kontzer goes on to make the point that what is important for organisations is 'what individual workers do with the knowledge they have'. When employees leave a company, he says, 'their knowledge goes with them, no matter how much they've shared' (Kontzer 2001). We cannot package it and put it on a shelf, ready to hand to the next person who walks in.

Wilson (2002) reiterates this point in his argument for the personalised nature of knowledge when he says that "whenever we wish to express what we know, we can only do so by uttering messages of one kind or another - oral, written, graphic, gestural or even through 'body language'." He argues that these latter communications do not carry 'knowledge', but constitute 'information'. When we communicate information, we rely on our listener assimilating our shared information into their own schema of knowledge. Each individual's schema of knowledge is different, because it is, as Schutz (1967) would say, 'biographically determined', therefore a piece of 'knowledge' (derived from a set of information) is never the same for any two individuals.

This brings us to the issue of deliberate misapplication of knowledge by employees in a business. In their book 'Organisational Misbehaviour' Ackroyd and Thompson (1999) show that misbehaviour is endemic in organisations. Though management frequently tries to control and eliminate misbehaviour, such practices are deeply rooted in the formation of employee interests and identities, and in the informal organisation developed to promote and protect them.

Could it be that in conventional KM we find ourselves on a path that is ultimately impossible? The goals of KM, as defined previously by Barth (exploiting intellectual capital to gain competitive advantage) however are as pertinent as ever. So perhaps there is an alternative that is effective and efficient.

Our case study focuses on an organisation that has concentrates on measuring achievement (outcomes). When we talk about outcomes we are talking about implied and applied knowledge, as knowledge must be drawn upon in order to create an outcome. Our case study examines what outcomes one organisation has chosen to measure and how those measures change the behaviours of the individuals who work there.

3. Case study

The data used for this case study is the ethnographic data from research conducted within a small engineering company on the south coast of England. This company has recently changed its management structure from a traditional top-down hierarchical organisation (dendritic, functional, command and control model) to a flat-structured enterprise with self-managing, autonomous individuals managing their own projects from start to finish.

The individuals are responsible for everything from securing future work flow, through design, manufacture and despatch of goods, to final invoicing and archiving. All departments and functions within the business were disbanded so each individual is now responsible for all aspects of the job and its management.

The point must be emphasised that the organisation has not created self managing autonomous work teams, but rather deconstructed the business down to the level of multi-skilled individuals. Their desire for autonomy and self-organisation is exploited within a systemic framework of measures that identifies and measures customer satisfaction and economic value-add.

Whilst this is at odds with conventional leadership wisdom and the division of labour, this approach has provided benefits (customer satisfaction and economic value-add).

In a paradigm where specialisation is promoted from an early stage (in the UK it begins while we are still at school), it seems strange to outsiders that this company requires its employees to multi-skill and be competent salesmen while also excellent design engineers, project managers and accountants. One comment has been 'isn't that the job of ten people?' illustrating that in our wider culture we expect not to be multi-skilled at work. However, the individuals who work in this company now are, on the whole, managing their work more successfully – and that success is measured by higher customer satisfaction due to better quality, delivery records and competitive prices – than when the business ran a conventional management system.

The model underpinning the organisation's new system is best understood from the perspective that businesses demand dynamic compromises to best respond to circumstances that present themselves; a multi-skilled person is better able to make cost effective compromises than a group or team effort (who have only static prescriptive solutions at their disposal). The essential characteristic of the company's new model is that the nature of individual responsibility has changed from a model where responsibility is distributed to one where it is focussed.

Consider the range of people who contribute to the production of a part. In a conventional model, many people contribute their competencies to the completed part, they are co-ordinated by managers, foremen and IT (and all of these are a cost).

In this business approach very few people, perhaps only one individual contributes to the part, and there is only ever one individual who is *responsible* for the part.

The benefits of the division of labour are lost, but the costs of the co-ordination are saved and customer satisfaction has improved (as has *employee* satisfaction, but that is beyond the scope of this work).

The changes to this organisation were introduced step by step, and here we will outline some of the most prominent.

- Removal of overtime.
- Distribution of centralised stores to stock kept where it is used by an individual
- Personal stock management
- Personal purchasing and reduction in purchasing costs
- Personal goods inward approval
- Removal of clocking-in machine and time cards
- Tracking of customer returns back to individual
- Personal commitment of delivery to customer, and tracking of performance.
- Introduction of an internal market
- Personal creation of inspection criteria
- Personal packaging and dispatch of goods
- Personal order processing
- Personal Profit and Loss account plus balance sheet.

- Attribution of product sales income to an individual
- Introduction of profit related bonus (20% of profit achieved by an individual pcm)

With the removal of management, the cost of running the organisation reduced. Simultaneously the productivity of the workforce increased. Where before individuals managed their work in such a way that it would warrant overtime, so that they could earn more money, the new system rewards good quality and delivery performance (these are measured on each individual every month). There is now the situation where each individual can take home 20% of any profit he or she makes in a month. So by measuring the profit they make, profit making behaviour is indirectly improved.

Similarly with 'knowledge': there are a variety of skills and knowledge that each individual needs in order to successfully manage a project. In order to improve the skills and knowledge of an individual, their outputs must be measured.

This leads us to the issue of engagement. To engage someone is to hold fast a persons attention (OED). Restructuring the organisation on an individual basis has indirectly increased the level of engagement demonstrated by the employees. It is impossible to function successfully within the business without a high degree of genuine engagement, because to produce successful outcomes an individual must use their skills and knowledge. Within this engagement a person brings their knowledge to bear.

We observe that individuals will search out answers to problems when in doubt; they choose to fill holes in their knowledge when these holes become apparent. They test, experiment, research and enquire. Informal communication provides for the cross pollination of best practice within the business and the personal interpretation of this best practice provides for constant reaffirmation of its validity.

As the individuals who work in this company have acclimated to their positions of multi-skilled managers of their own projects, the company results for the measurement of Quality and Delivery have improved. (see figures 1 and 2.) The directors of the company had to take the risk of letting go of their control over what the individuals did to achieve their goals, even though it seemed counter-intuitive. Yet the results tell their own story. The individuals now do a better job:

Figure 1: Quality perfomance, as measured by the % of unreturned parts

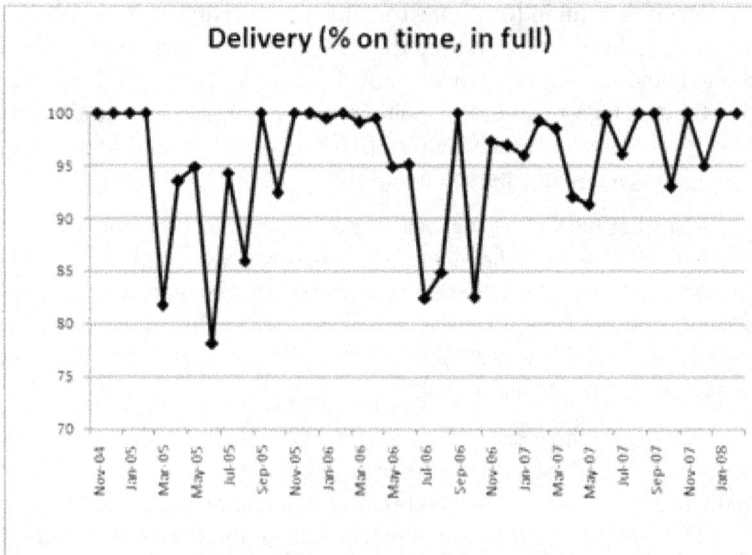

Figure 2: Delivery performance, as measured by % of deliveries on time, in full.

Both these graphs illustrate the improvement in performance over time of the employees.

This case study shows that the change of focus from the micro-management of people and what they know and do, to the management of their performance in terms of outcomes has improved their business performance and the change of approach has proved to be a worthwhile investment.

4. A different approach to KM

The technique used in this organisation has been to create systems and a culture in which people need knowledge to prosper. The organisation is one in which personal interdependence is such that the sharing of knowl-edge provides a tangible benefit, then knowledge management is precipi-tated; and demonstrably so within the outcomes of the business.

Conventional KM recognises this need in organisations, but many try to artificially induce extra knowledge as some extrinsic dimension that can be added on. If we examine closely Polanyi's theories of knowledge, he states that 'certain cognitive processes and/or behaviors are undergirded by op-erations inaccessible to consciousness' (Barbiero, n.d., cited by Wilson 2002).

Thus the point here is that what he refers to as 'tacit' knowledge means something that is 'hidden', hidden even from the consciousness of the knower (Wilson 2002). Polanyi (1958) used the phrase 'We know more than we can tell' precisely to communicate the inaccessible nature of per-sonal knowledge, inaccessible even to the consciousness of the knower. That tacit knowledge can somehow be 'captured', as is the claim of Nonaka (1991) and Nonaka and Takeuchi (1995), is a puzzle. As Wilson asserts, 'tacit knowledge is an inexpressible process that enables an assessment of phenomena in the course of becoming knowledgeable about the world... it cannot be captured – it can only be demonstrated through our expressible knowledge and through our acts' (Wilson 2002).

Thus this organisation uses an alternative approach which is to work in alignment with Coverdale's (1977) values, and to focus upon the 'doing' skills and knowledge of an individual rather than knowledge as an extrinsic entity, as these kinds of skills are the tangible outcome of knowledge use.

Coverdale links difficulties, knowledge, actions, outcomes and risk together: he shows us that when we meet a difficulty we must bring our knowledge to bear on the problem (he calls these synthesis and analysis) and act (do something, use skills) and take risks (we don't know if it will work out) to produce an outcome which is a resolution to the problem.

Risk is present at both personal and corporate levels, the purpose of KM is to ameliorate the risk from the corporate perspective. It is assumed that to develop a KM strategy is to protect this intellectual capital by making it tangible and to preserve it once the individual who carried it leaves.

Companies, especially in the service industry, rely upon the knowledge of their individual's for their income. As Barth (2002) points out, 90% of their assets (intellectual capital) are carried around in the heads of the workforce.

This is a paradox, because at the heart of conventional KM is the assumption that knowledge can exist separately from the individual who carries it, and through separation risk is reduced. But the act of separation demotes the knowledge to the level of information and actually increases risk. This may be the reason that outcomes in KM programs are inconsistent.

Here then is the heart of our difference in approache. Where conventional KM has tried to separate knowledge from the individual systemically, our approach promotes knowledge as an inextricable part of the individual and provides both responsibility and risk at the level of the individual.

During our practical exploration from within a real business environment, we believe we have determined that knowledge can not be transferred explicitly and completely (identified, transferred, collected, stored, retrieved and applied) in a *cost effective* manner. We are not suggesting knowledge cannot be managed, but rather that doing so is not cost effective; it may even be cost prohibitive and culturally damaging.

We believe that KM follows the law of unintended consequences; it is always incomplete, and the work required to complete it introduces more problems. And whilst more effort is put into completing it, completion is never reached. There are always more instructions or hints and tips one could add. Furthermore, the cost of this effort applied to all the knowledge put into storage is not repaid by the successful application of the subset of knowledge which is actually retrieved.

We explored many alternative ideas whilst developing this work, including the idea that managing even highly incomplete knowledge can provide value and an investment return. But whilst this looked promising, we predicted a significant problem with getting operators to use a "library" with knowingly incomplete knowledge in it; uptake being crucial to the return on the investment.

For the KM process to be complete it has to include the following: the development of new knowledge (hypothesis, experimentation, application and affirmation), its capture (condensation) and transformation (communication), collection (accumulation) and categorisation (to make it searchable), its storage (secure and accessible), retrieval, then comparison of the original context to today's circumstances, the prioritization of all the knowledge revealed in the search, and the application of knowledge in pursuit of a commercial purpose (quality, delivery, profit or conformity), and the final realization of the commercial goal.

The more useful and explicit one makes knowledge in preparation for storage, the more expensive is its management. Expensive KM is not in itself a problem, however the investment return on that expense is. How much is the knowledge worth? Each bit of explicit knowledge has only a finite value, and that value is determined by the circumstances of the future, and the successful application of this knowledge in those circumstances. The organisation started to search for a more useful way of viewing knowledge. In hindsight, the shift was made to view knowledge as a dynamic and infinite abstraction, and therefore not treat it as a "real" thing; not to *reify* it.

Returning to the practical exploration of the KM problem, it remained the organisation's goal to achieve the commercial benefits of KM; they were struggling to engineer an investment return on their KM program, yet it remained central to the commercial success of the business. It was a barrier that had to be successfully overcome.

The solution is to pioneer an alternative approach that provides the commercial benefits of KM, but without the explicit management of knowledge. The experience of the business has been that the secret to commercial success of KM is its IMPLICT management, in contrast to explicit management. And that explicit KM does not provide a net investment return, it is unreliable and culturally counter productive.

In pursuit of communicating this approach, knowledge (K) can be viewed as an *abstraction*, or a first order derivative. It is something that cannot be directly observed, rather its effects can be observed. Having found this derivative (K) very difficult to manage explicitly, a *consequence* of the knowledge was sought, that might provide something real to manage.

Thus their attention was turned to the "outcome" of K, specifically, to find something that would imply the use of managing K (like a shadow implies a solid object). It is reasoned that if the organisation can reliably precipitate that outcome, then the KM it implies must be occurring as well.

The organisation focuses on identifying and rewarding the individuals within the organisation who overcome the commercial barriers of the present day. Those who can satisfy their customers profitably, even in tricky circumstances, are rewarded each month. When this second order derivative (commercial outcome) is managed, it precipitates KM. Furthermore it tends to elicit very unconventional KM techniques by its operators. Slick, fuzzy (logic), innovative, even at times counter intuitive, these techniques are providing commercial outcomes and clearly implied knowledge management, but are not fixed and deterministic in nature. Through managing the consequences of achieving commercial outcomes, the organisation is implicitly inducing a fluid KM strategy. And the KM that is elicited this way is diverse, yet successful. For example, can you get your competitor to unknowingly, yet securely store knowledge for you at their expense? It may seem like a strange thing to do, but it is a successful KM strategy.

Knowledge can be usefully viewed as a collection of *cultural memes*. This idea goes some way to address further complications in KM that were encountered, for example the context and nature of the stored knowledge was constantly moving, not just during retrieval but also within the "static" storage phase. Rather than knowledge repositories being a dusty storage of bits of data and facts, every new piece of knowledge was re-writing the old previously stored knowledge and making it obsolete or newly applicable.

Knowledge, it was found, is not fixed, it is dynamic. It is advantageous to treat knowledge as changeable of the pursuit of commercial success in the current circumstances, rather than the replaying of previous knowledge schema. There is a tendency to reify it; and when we tend to treat the abstract as real in error, and we suffer accordingly.

This idea of knowledge as a cultural meme provides the opportunity to think about the diversity of solutions that may be applicable to a particular problem. That is to say, presented with a particular problem there are a number of approaches that could be taken to reach a solution, based upon different knowledge routes. This diversity of knowledge provides a richness of choice to the solution set, it is not about identifying the knowledge to solve the problem, but about which solution gives you competitive advantage. As promoted by Ashby's Law of Requisite Variety, a range of solutions is only available through diversity.

We believe that there is cultural risk in turning knowledge management into a process where knowledge is reified, because this destroys the ability of the future user of that knowledge to overcome the unique barriers of their circumstances. There may be short term benefits of neatly packaged KM, but there is a clear danger of creating a knowledge 'mono culture'. The aim must be to have both the short term gains of effective KM, plus the long term gain of a diverse knowledge culture.

If knowledge is viewed as a collection of cultural memes, destroying cultural meme diversity is a high risk strategy. For example, distil the genetic diversity of a crop and yields improve, but the long term risk is high because a day will come when the entire crop is decimated by some hidden vulnerability. It is genetic diversity, and subsequent exponential growth of the surviving strains that provide your only hedge against complete crop failure and subsequent famine.

The diversity of cultural memes is key to ensuring a business is capable of tackling the unique commercial barriers that exist in its future. Rather than treating knowledge as something to fix and distil, it must be nurtured and diversified. Thus we believe this approach to KM uniquely satisfies both short term goals and longer term risk amelioration.

Returning to our idea that knowledge lies within the individual, and to Follett's definition of management as 'the art of getting things done through people' (Barrett 2003:51), the anthropological concept of agency further informs the philosophy behind our argument that the focus of knowledge must lie within the individual. The concept of agency is developed by Nigel Rapport (2003) as indicative of an individual's capacity for action in the world. Rapport illustrates his argument with case studies of four individuals whose lives were characterised by taking unusual paths, trajectories he called them; ones that seemed at odds with the hegemonic ideals, and took strength of character to uphold. In this instance we would advocate agency as something each person possesses that allows them to act independently of instruction to meet their needs.

When the individual is made responsible for their own risks in an effort to achieve outcomes, they will, through their own agency, bring their own dynamic knowledge set to bear on the situation. Through their engagement with the problem, knowledge is used, skills are drawn upon and solutions found from wherever they need to come from. There are many more possible solutions to a problem when an individual has the freedom to find them than when management prescribes a process for problem solving.

Our argument is that when conventional management is applied to knowledge, knowledge becomes reified as something extrinsic to individuals. This is not the true solution to managing knowledge for 'exploiting intellectual capital to gain competitive advantage', as our case study and argument shows.

Therefore knowledge needs to be approached as something non-manageable. Something that is inextricable from the individuals within the business, and that the best way to 'manage it' (that is, get the best use out of it) is to allow the individual to manage it themselves.

This paper does not recommend the outright replacement of conventional KM, but rather serves as a reminder that KM can only be cost effective when in the context of systems, measures and a culture that is not dysfunctional.

References

Ackroyd, S. and Thompson, P. (1999) Organizational Misbehaviour London:Sage.

Barrett, R. (2003) Vocational Business: Training, Developing and Motivating People. Cheltenham: Nelson Thornes.

Barth, S. (2000) Defining Knowledge Management, definingCRM online magazine [http://www.destinationcrm.com/articles/default.asp?ArticleID=1400 accesed 3/4/08]

Drucker, P.F. (1969) The age of discontinuity: guidelines to our changing society. New York, NY: Harper and Row

Foyal, H. (1967) General and Industrial Management. (trans. from the French ed. Dunod) by Constnace Storrs, with a forward by Urwick.L. London: Pitman

King, W.R; Chung, R; Haney,M. (2008) Knowledge Management and Organizational Learning OMEGA; 36(2): 167-172.

Kjaeraard, A; Kautz, K. (2008) Process Model of Establishing Knowledge Management: Insights from a Longitudinal Field Study, OMEGA; 36(2): 282-297

Kontzer, T. (2001) Management legend: trust never goes out of style. Call Center Magazine. http://www.callcentermagazine.com/article/IWK20010604S0011 [accessed 3/4/08]

Nonaka, I. & Takeuchi, H. (1995) The knowledge creating company: how Japanese companies create the dynasties of innovation. Oxford: Oxford University Press.

Orzano, A. John; McInerney, Claire R.; Tallia, Alfred F.; Scharf, Davida; Crabtree, Benjamin F. (2008) Family medicine practice performance and knowledge management. Health Care Management Review 33(1):21-28.

Polanyi, M. (1958) Personal knowledge: towards a post-critical philosophy. Chicago, IL: University of Chicago Press.

Schneble, John (2003, August). The Need to Know. Informationweek. [www.informationweek.com/story/showarticle.jhtml?articleid=13100330 accessed 2/4/08]

Schutz, A. (1967) The phenomenology of the social world. Evanston, IL: Northwestern University Press.

Smith, M. K. (2003) 'Michael Polanyi and tacit knowledge', the encyclopedia of informal education, [www.infed.org/thinkers/polanyi.htm accessed 2/4/08]

Wilson, T. D (2002) the nonsense of 'knowledge management' Information Research: an electronic journal, 8(1) [http://informationr.net/ir/8-1/paper144.html, accessed 3/4/08]

How to Design for Strategic Innovation? Appropriate Forms of Ambidexterity

Liselore Berghman

VU University Amsterdam, The Netherlands

First published in The Proceedings of ECIE 2009

Editorial Commentary

As previously noted, the growing changes in the economic environment oblige organizations to respond with innovation and entrepreneurship. They must create and profit from new business models (Govidarajan & Trimble, 2005). Strategic innovation implies rethinking the underlying logic of the business and making new choices regarding structure, staff, systems and culture, what Govidarajan & Trimble (2005, p. 48) term, "Organizational DNA".

Berghman investigates the structural design capabilities for strategic innovations, ie. capabilities required for the systematic creation of strategic innovations. She focuses on the concept of *ambidexterity* which connotes the organization's capacity to systematically balance exploration and exploitation. She explains to which extent the ambidexterity approach should be attuned to each contextual situation as well as to each stage of the innovation project. This nicely contributes to the academic literature by feeding a nascent research stream of *ambidexterity* with implications for innovation: developing new insights on structural variables that benefit a particular type of innovation process.

References:

Govidarajan V. & Trimble C, 2005. Organizational DNA for strategic innovation, California Management review, vol.47 N° 3

Abstract: In this paper we study one specific type of non-technological innovation: strategic innovation. Strategic innovation follows a Schumpeterian perspective entailing a deviation from the traditional competitive rules of the game with the aim of offering a new and superior value proposition. As companies creating strategic innovations show high revenue and profit growth a deeper understanding of the underlying organizational capabilities would be highly relevant to both researchers and managers. Therefore, and in contrast to prevailing research on strategic innovation, we do not study the phenomenon on an industry level but study the capabilities that a business unit needs for the systematic creation of strategic innovations. More specifically, we study structural design capabilities and focus on the concept of 'ambidexterity'. The central theme in this nascent research stream is which (if any) organizational design allows firms to be efficient and adaptive/innovative at the same time. We use a two-phase qualitative design to study business units' systematic creation processes of strategic innovations in five Dutch industrial sectors. Our findings add to the emerging academic discussion on ambidexterity by showing that the appropriate ambidexterity approach may not only differ by innovation type but also by the specific phase of the innovation project. Our data show that companies have to cope with different ambidexterity frictions in the initiation and commercialization phases of strategic innovation projects. More specifically, the initiation phase rests on 'contextual ambidexterity': the initiation of strategic innovation initiatives is interwoven with the normal core business operations. To this end, companies establish mechanisms that stimulate the recognition and assimilation of customer/market knowledge. In contrast, 'structural ambidexterity' is applied during the commercialization phase, meaning that strategic innovations are gradually separated from the core business. These findings are relevant to managerial practice as they suggest a limited value of isolated business development units for strategic innovation initiation.

Keywords: Strategic innovation, structural design, ambidexterity

1. Introduction

Although innovation has been a popular research theme in diverse academic fields (Wolfe, 1994, Gopalakrishnan & Damanpour, 1997) studies have primarily focused on technological and product innovation types. The growing attention for non-technological innovation types in policy programs has not been fully reflected in academic research (Birkinshaw et al.,

2008, Siguaw et al., 2006). Therefore, this paper focuses on one type of non-technological innovation: strategic innovation (SI).

In markets characterized by a near-perfect competition state and advanced commoditization levels companies can still achieve above-normal profits by looking for fundamentally new ways to differentiate themselves (Larsen et al., 2002, D'Aveni, 1999). This so-called 'strategic innovation' (SI) follows a Schumpeterian perspective, focusing on innovation of the business model, offering a new and superior value proposition and breaking with industry rules of competition (Govindarajan & Trimble, 2005, Kim & Mauborgne, 1999, Markides, 2006).

Research on SI has its origins in more popular management literature (Markides, 1997, Kim & Mauborgne, 1997) but has gradually received more academic attention (eg., Govindarajan & Kopalle, 2006). Scientific contributions have tackled the phenomenon primarily on an industry and less so on an organizational level of analysis (Markides, 2006, Larsen et al., 2003). Yet, as companies creating SIs show high growth (Kim & Mauborgne, 1999, 1997) a deeper understanding of underlying organizational (design) capabilities would be highly relevant to both researchers and managers.

Therefore, we study structural designs for SI. More specifically, we focus on 'ambidexterity': the organizational design capability to systematically balance exploration and exploitation (Gibson and Birkinshaw, 2004, Tushman & O'Reilly, 1996). To this end, we study qualitative data of companies in five Dutch industrial sectors. Our analysis suggests that companies apply different ambidexterity strategies in the initiation and commercialization phases of SIs.

Our study aims to answer recent calls for more research on structural variables for innovation (Hauser et al., 2006, O'Connor, 2008, O'Connor and DeMartino, 2006). We try to extend research on SI in particular, by illuminating micro-level —instead of industry-level— implications. Finally, our study responds to calls for a more 'granular view' on ambidexterity research, where implications for different types of innovations are better specified (Raisch & Birkinshaw, 2008).

To build our argument, we first introduce the notion of ambidexterity and review the literature on structural designs for SI. After explaining the research method we will present and discuss the findings of our analysis and develop some propositions. The paper concludes with theoretical and managerial implications.

2. Structural designs for strategic innovation

The seminal article by March (1991) on the self-reinforcing trade-offs be-tween exploitation and exploration has spurred research on ambidexterity. The central theme in this nascent research stream is which organizational design allows firms to be efficient and adaptive/innovative simultaneously. Overall, ambidexterity research considers two different strategies to man-age innovations (Raisch & Birkinshaw, 2008). The *structural ambidexterity* strategy builds on Duncan's (1976) argument that organizations need mechanistic structures to exploit innovations but need organic ones for creating them. The tactic of spatial separation comprises to create and launch innovation initiatives in separated stand-alone units (eg., Tushman & O'Reilly, 1996, Christensen et al., 2002). In contrast, *contextual ambidex-terity* allows for simultaneous exploration and exploitation in the organiza-tional core through the creation of a specific 'enabling' organizational con-text (Gibson and Birkinshaw, 2004, Adler et al., 1999).

Research about when and how to make effective use of each strategy is however limited (Siggelkow & Levinthal, 2003). Although it is widely recog-nized that different types of innovation require different (structural) condi-tions (McDermott & O'Connor, 2002, Abernathy & Clark, 1985) these dif-ferences have surprisingly not been integrated in ambidexterity research (Raisch & Birkinshaw, 2008). Therefore, we reviewed recent strategic inno-vation studies on the structural designs that are proposed (see Table 1). Proposed designs essentially pivot on arguments of a) maximizing syner-gies by totally integrating the new unit within the existing organization, or b) minimizing conflict and risk by full separation. Integration strategies build on the principle of contextual ambidexterity while separation strate-gies rest on structural ambidexterity. Table 1 shows that by and large three different structural designs can be discerned.

Among the adherents of the *integration* approach are Kodama (2003), Hamel & Getz (2004) and Iansiti et al. (2003).

Table 1: Structural designs for strategic innovation

Author, year	Research	Key findings	Proposed structural design for SI	Ambidexterity strategy
Kodama, 2003	Qualitative study of SI process in two large Japanese companies	Inclusion of paradoxical elements throughout entire organization by means of knowledge management and external alliances	Integration	Contextual
Ansiti McFarlan & Westerman, 2003	Large multi-method study of more than 100 high-tech, retail and non-retail firms	(Re-) integration required for long term success (synergies with parent). Three potential integration strategies	Integration (Full or phased)	Contextual (or loosely structural)
Hamel & Getz, 2004	Conceptual study based on examples of hyper-efficient innovating companies	Invest in employees' capacity to innovate innovation as deep value, strategic experimentation,...	Integration	Contextual
Christensen & Overdorf, 2000	Conceptual study based on examples of disruptive innovations in established companies	Heavyweight team ir separate spin-out organization best fit for innovations with poor fit to organizational values (even if fit with organizational processes)	Separation	Structural
O'Reilly & Tushman, 2004	Qualitative study of 35 breakthrough initiatives undertaken by 15 BUs in 9 industries	Separate exploratory units with tight integration at senior management level. Separation allows for unique processes, structure and culture for innovation units	Separation	Structural
Govindarajan & Trimble, 2005	Qualitative, longitudinal research of best practices in the management of strategic experiments in large established organizations	Experiments ir separate units with extensive organizational links to mother company enable forgetting, borrowing, and learning between old and new business	Loose separation	Structural
Raisch, 2008	Qualitative study of the use of temporal separation, structural separation and parallel designs in 6 central European companies	Separate creation of new growth platforms require high level of autonomy (idea generation & capabilities: own value chain independent (brand name) but also many operational links for efficiency (nurturing through specialist support & sharing of assets)	Loose separation	Structural
O'Connor, 2008	Conceptual study	Dedicated group with both tight and loose interface mechanisms to mainstream organization	Separation / Loose separation	Structural
Markides & Charitou, 2004	Multi-method study of 10 case studies of new business model introduction in established companies + large-scale survey	Four potential strategies dependent on a) degree of conflict with established business model and b) degree of relatedness of markets	Four possibilities: •Separation •Phased separation •Phased integration •Integration	Structural (different forms over time)

Proponents of *structural separation* all depart from the difficulties associated with the incorporation of radically new ventures in existing operations (Burgelman, 1984, Christensen & Overdorf, 2000, Christensen et al., 2002). Christensen & Overdorf (2000) and O'Reilly & Tushman (2004) take the most radical stance whereas more recent studies by Govindarajan & Trimble (2005), Raisch (2008) and O'Connor (2008) have taken a more nuanced position. The latter are in favour of a so-called *'loose separation'* strategy, where the separate innovation units maintain extensive links to the mother company in order to also leverage existing assets and capabilities. According to Govindarajan & Trimble (2005) such a 'dual-purpose organization' possesses the ideal characteristics for SI.

Finally, Markides & Charitou (2004) propose four different designs that range from total separation, over phased separation, phased integration to a total integration strategy; the appropriateness of each strategy depends on a high/low degree of conflict between the new and established business model and the markets targeted. However, many SIs can not be classified as one of these extreme cases and typically involve both a deviation from the existing business and a leverage of existing assets and capabilities.

In conclusion, even though the antinomy between integration and separation strategies has become softened, Table 1 shows that still no strong evidence exists about *the* appropriate ambidexterity approach for SI. Furthermore, different innovation phases (eg., Cooper, 2008) represent different challenges and require different competencies (O'Connor, 2008, Hurley & Hult, 1998). Although some authors' findings suggest a different structural design throughout the different stages of SI creation (Iansiti et al., 2003, Markides & Charitou, 2004), none of the studies examined in Table 1 has incorporated this aspect explicitly. In addition, the temporal aspect of innovation processes has not yet been dealt with in ambidexterity research either (Gupta et al., 2006).

3. Method

A qualitative research design was consciously chosen as we judged theoretical insights as too premature to be immediately subject to theory testing. Qualitative research is particularly useful in theory development and refinement (Bacharach, 1989).

The empirical research centered on a two-phase qualitative research design (see Figure 1).

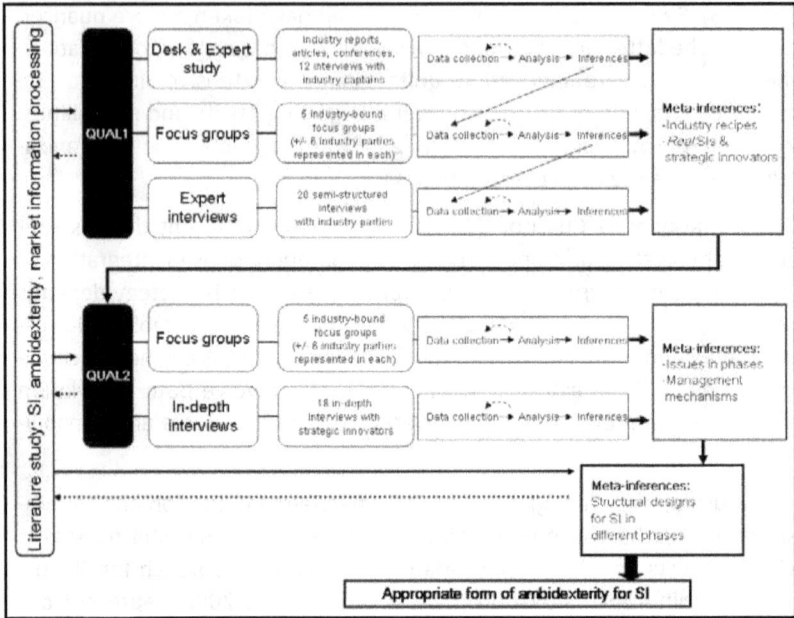

Figure 1: Research design

The use of different qualitative techniques in each phase enabled be-tween-method triangulation (Tashakkori & Teddlie, 1998). The full and dashed arrows in Figure 1 show furthermore how we iterated between data collection and analysis (Eisenhardt, 1989) and between theory and empirics (Orton, 1997).

The definition of SI implies contrasting SI initiatives with the competitive rules of the game in an industry. We hence selected 5 industrial sectors (traffic management, energy, food, truck&trailer, graphics printing) and studied the industry rules of the game by means of different methods. We studied secondary data and interviewed different key informants to each industry. To validate findings, a focus group for each industry was then set up. Each focus group had 6 participants on average, all representatives (mostly CEOs and marketing managers) of both general players and niche players in the industry. Finally, we conducted 28 individual interviews with industry parties to refine the 'industry recipes' (see also, Porac et al., 1989, Matthyssens & Vandenbempt, 2003, Spender, 1989) and to select 'real' SIs in each sector. After selecting SIs, we controlled for SIs as 'lucky shots' (Govindarajan & Trimble, 2004) and selected firms/BUs with the capacity to create SIs *systematically*. These 'strategic innovators' are firms/BUs that had launched several real SIs.

For the second qualitative phase we followed extant research and shifted the level of analysis to the BU (Mizik & Jacobson, 2003, Jaworski et al., 2000). Focus groups were still industry-bound and consisted of (BU, marketing) managers, representing companies with a high or low level of strategic innovation capacity. The focus groups tackled the issues managers were confronted with during the creation of SIs and the specific mechanisms to manage these. Then, we held 18 interviews with managers who were highly involved in the set-up of SIs to validate findings. We followed Lynn et al.'s (1996) argument to study successful cases in diverse contexts rather than to compare successes and failures in similar contexts.

Miles & Huberman's (1994) procedures were applied for data coding and analysis. We bundled stretches of data along several content categories. For example, for the second qualitative phase we analyzed data according to the internal-external and the initiation-commercialization dimensions. For the first phase, a content analysis was performed per industry. We furthermore structured the data stretches by source and looked for (dis)similarities across the different sources per industry. For the second phase we applied within-method triangulation and triangulated interview data with organizational documents or with additional interviews (with customers or other parties involved in the SI).

4. Results

4.1. Issues in the creation process of SI

The qualitative data yielded different barriers that companies are confronted with throughout the creation process of SIs. Table 2 shows the bottlenecks mentioned during the initiation and commercialization phase (Gopalakrishnan & Damanpour, 1997).

Barriers in the initiation phase are more of a cognitive nature, whereas companies essentially struggle with resource allocation, political and market issues in the commercialization phase (cfr., McDermott & O'Connor, 2002).

Table 2: Issues in the initiation and commercialization of SI

INITIATION		COMMERCIALIZATION	
Internal	**External**	**Internal**	**External**
Limited and biased customer insight Narrow market insight Initiators of SI ≠ possessors of customer/market knowledge No time for critical reflection on customer/market assumptions/information No cross-functional discussions on customer/market assumptions/information	Limited information from customers	Chaos & long-winded process Interference with current operations Combination of new & existing competencies	Current market image limits market acceptance of SIs Hostile industry climate (acts of revenge)

Internal issues in the initiation phase center on firms' abilities to recognize and assimilate market knowledge. A printer says *"you should develop deep customer insight, and know what the market is like, now and in the future. It takes an awful lot of energy to deeply think about this, to think beyond your traditional business. But it is crucial, very crucial"*. Furthermore, all strategic innovators indicate that the wealth of customer and market information sales people dispose of should be disseminated across the business unit. The marketing department was somewhat marginalized in this respect. *"A marketing department is good at 'packaging' new concepts, but not at inventing them. They are too far away from the market, too theoretical, too model-driven"* (Energy company). A strategic innovator in the truck & trailer industry notes: *"Time to cross-functionally reflect, especially on market and customer information, is crucial. It is an awkward and time-consuming process to find the right formula for this. Especially SMEs do often not find the energy to put a systematic effort in this activity"*.

Regarding external issues during initiation, strategic innovators highly value customers' information about market developments, competitors and potential industry entrants, which helps developing new SI ideas and predicting market and industry reactions. A hardware producer in the graphics printing industry remarks: *"the fact that we are innovative is for sure dependent on our own innovation capacity, but, above all, we have good customers"*.

Concerning the commercialization phase, strategic innovators strongly agree that chaos should be avoided because it worsens market credibility and considerably retards implementation. Many interviewees spontaneously bring up that commercialization is a long-winded process. In addition, most of them talk about taking smart and calculated risks. A truck & trailer supplier says: *"you need SI to escape price competition. But next to this, you still need the traditional business. It's just a matter of not putting all eggs in one basket"*. Finally, the combination of existing and (sometimes) company-foreign competencies is sometimes hard to manage. The incompatibility in terms of market image, is an often-cited problem as well. For example, a producer of printing hardware mentions the difficulty to profile itself as an independent full service provider, while traditionally producing and selling one specific hardware brand. In particular, strategic innovators warn against brand interference. The data furthermore suggest that a firm's SI capacity is also stimulated or curbed by the general climate of the supply chain. Several strategic innovators remark that the traditional market has become so small and commoditized that the majority of incumbents fear that *"one man's breath is the other one's death"*. Interviewees assert that many parties are 'expectant' or even reluctant to take up SI, out of fear of acts of revenge by other industry parties.

4.2. Effective management mechanisms in the creation process of SI

A study of the way strategic innovators avoid/overcome the barriers shown in Table 1 provides insight into effective management mechanisms for SI creation. Overall, a pattern arises where strategic innovators establish deliberate mechanisms to overcome cognitive inertia during the initiation phase.

For the initiation of SIs the need to deeply study *existing* customers and to build a 1:1 learning relationship with them is highly stressed. Strategic innovators share the view that personal, deep customer relationships increase a proactive market interpretation. For example, a food ingredients supplier remarked: *"it [the creation of a SI] is not reacting to a crisis, neither is it pure proactiveness. Instead, it is a sort of logical implication of carefully listening to your customers and mapping their needs to your own competences"*. Non-sales people take a lot of time to join account managers on customer visits and to talk with users. However, emphasis is put on the most innovative customers to acquire innovative ideas. For example,

91

many firms indicate they use special user groups, where they invite a limited number of innovative customers to jointly discuss their problems and suggested solutions. Large-scale market research is strongly rejected.

In line with the literature on SI (Markides, 1997) strategic innovators stress the responsibility of the entire organization for recognition activities: *"During years, service was considered the necessary evil, now it is our market information antenna"* (graphics printing systems supplier). Often, recognition aspects are incorporated within performance appraisal criteria of sales and maintenance people.

Furthermore, the value of internal discussions about customers and the market is highly stressed (cfr. Tripsas & Gavetti, 2000). Sometimes, the entire industry and supply chain is being discussed and topics such as: downstream interference, the definition and demarcation of the business, the exclusivity issue of SI and its incompatibility with the traditional business are dealt with.

Often, discussions take place as periodical, cross-functional meetings (marketing, sales, BU management) but sometimes external parties such as customers, partner-companies, or university professors are involved as well. Strategic innovators indicate that different perspectives deepen discussions and help develop market scenarios for the future. A printer remarks: *"Internal soundboards are extremely important. They let you think, hear other visions. They keep you 'on edge'"*.

Concerning the commercialization phase the data uncovered a totally different pattern. Almost all SIs imply an adaptation of the organizational structure. Often detached temporary project teams are set up, that take care of marketing and commercial aspects; for other issues they hire resources from the existing units. Afterwards, project teams grow out into fully separate units. Interviewees mention that this approach limits and smoothes a sudden, drastic structural adaptation. The use of a strict project-driven approach and the set-up of separate units are mentioned as highly valuable tools to cope with chaos.

Strategic innovators often use new hires, experienced in different industries, to fill competence gaps For example, a truck & trailer supplier hired people from the insurance sector. These new hires are mixed with existing employees, who are fully, part-time or temporarily dispatched to the new unit.

To avoid brand-interference (and cannibalization of the traditional business) SIs are marketed with a different brand name. Initially, even other
industry parties do hence not realize that the innovation is launched by a
well-known incumbent.

5. Discussion and conclusions

5.1. Discussion of the findings

The results of the qualitative study show the mechanisms that strategic
innovators establish to manage the different bottlenecks during the creation of SIs.

Concerning the initiation phase two categories of mechanisms can be discerned: a) mechanisms that stimulate the organization to better recognize
and acquire useful market/customer information and b) mechanisms that
stimulate to reflect on this information. In line with the literature on absorptive capacity (Cohen & Levinthal, 1990), we call them *mechanisms for
recognition* and *mechanisms for assimilation*.

The mechanisms for recognition confirm the importance of external
knowledge acquisition capabilities for proactive strategies (Tuominen et
al., 2004, Johnson et al., 2003). More specifically, findings provide empirical evidence for the 'proactive' market orientation perspective (Narver et
al., 2004) and take the edge off skeptics' warnings against the blinding effects of current customers and markets (see eg., Christensen & Bower,
1996). Firms with a proactive market orientation record and observe customers closely in their use of products or services (Slater & Narver, 1998).
Strategic innovators indeed deeply study the different stages in the buyer
experience process and the customer's own business processes to shed
light on new potential value propositions (Kim & Mauborgne, 2000). In
addition, the intensive consultation of innovative customers extends the
value of the lead user technique (Von Hippel, 1988, Lilien et al., 2002) to
the domain of SI.

The mechanisms for assimilation illustrate how strategic innovators do
invest in deliberate mechanisms to identify and discuss their implicit mental models of the market (Day, 2002). Our findings seem to support extant
sensemaking literature (Louis & Sutton, 1991), which has exemplified the
value of deliberate efforts for changing mental frameworks (Zollo & Winter, 2002). People with perspectives as diverse as possible are involved

(Markides, 1997) to constructively discuss long-held assumptions about customers and markets (Slater & Narver, 1995).

What is however more important in view of our research question is that the specifics of these mechanisms suggest that they largely depart from, and are therefore interwoven with the firm's daily operations. Or, the search and development (the 'initiation') of a SI tends to take place within the operating core of the organization whereas the commercialization is less so.

Therefore, we can propose that:

Proposition 1: The internal mechanisms that an organization establishes to develop a deep insight into (innovative) customers [*recognition mechanisms*] stimulate an organization's propensity to develop SI concepts.

Proposition 2: The internal mechanisms that an organization establishes to stimulate frequent cross-functional discussions about customer/market assumptions and information [*assimilation mechanisms*] stimulate an organization's propensity to develop SI concepts.

We know furthermore from research on sensemaking that overall, external scanning facilitates strategic action through its effects on strategic interpretation (Thomas et al., 1993).

Proposition 3: Mechanisms for assimilation will strengthen the positive effect that mechanisms for recognition produce on an organization's propensity to develop SI concepts.

In contrast with a contextual ambidexterity approach during initiation, a structural separation strategy (Benner & Tushman, 2003, Tushman & O'Reilly, 1996) is applied during commercialization. Separate SI-units are set up alongside traditional business departments to curb resource allocation conflicts and internal chaos (Burgelman, 1984). An important additional motive is however related to the management of *external* barriers to SI. The physical separation of SI units does not only make it easier to use different brand names but also to market the entire initiative in a 'hidden' way. This strategy helps in raising market acceptance and in avoiding hostile reactions from other industry parties. In order to leverage synergies with existing competencies (Volberda et al., 2001) 'switching' approaches (Adler et al., 1999) are additionally applied, where individuals are temporarily dispatched to SI-units or divide their responsibilities between both. In line with recent studies on structural designs for SI (Govindarajan & Trimble, 2005), a so-called 'loose separation' approach is followed.

Proposition 4: The application of a 'loose separation' strategy will stimulate an organization's capacity to commercialize Sis

Proposition 5: Industry hostility strengthens the positive effect of a 'loose separation' strategy on an organization's capacity to commercialize SIs.

Overall, our data suggest a phased separation pattern (Markides & Charitou, 2004). Figure 2 summarizes our findings and shows that the ambidexterity required for SI is likely to center on different, complementary strategies dependent on the specific phase of the SI creation process.

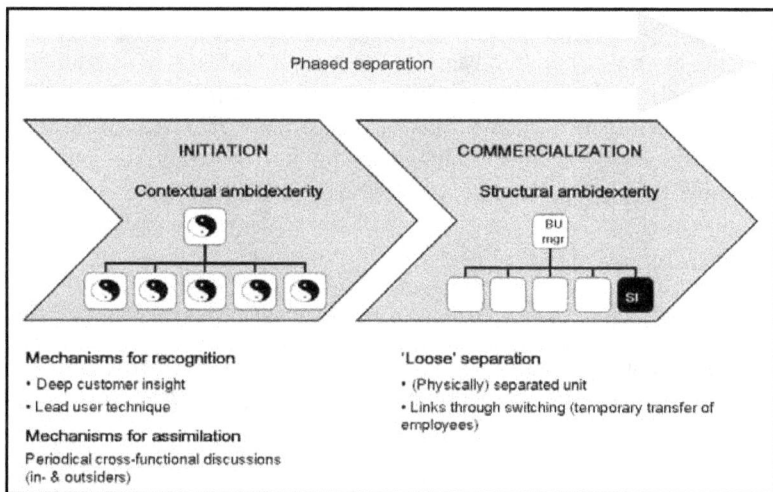

Figure 2: Ambidexterity forms for SI

5.2. Implications

This study adds to ambidexterity research by explicitly tailoring ambidexterity theory to one specific innovation type (Raisch & Birkinshaw, 2008). We believe this line of research can be extended towards other innovation types, such as studying whether strategic innovators apply the same ambidexterity approach for radical product innovation. Overall, research on innovation processes could serve as an input to future ambidexterity research and studies such as O'Connor (2008) and Griffin et al. (2009) would be excellent starting points.

Studying specific organizational management mechanisms we add to research on SI as well. Apart from validating our findings through a large-scale quantitative study, we think that integrating (interaction effects of) external variables (eg., customers' innovation attitudes) in our model could reveal potential trade-offs between internal and external market learning mechanisms for SI.

Our findings also have important implications for organizations that aim at creating SIs. Although innovation is always part serendipity, our findings show how SI can be stimulated deliberately as well. As SIs mostly develop through 'normal' business operations isolated business development departments are only of limited value to the initiation of SI. A new extended role of deeply studying customer needs is proposed for sales, marketing and even maintenance people. Finally, the commercialization of SI does not only require good internal management but also requires cautious management of external threats, such as hostile industry reactions.

References

Abernathy, W. J. and Clark, K. B. (1985) "Innovation: Mapping the winds of creative destruction", *Research Policy,* Vol 14, No 1, pp 3-22.

Adler, P. S., Goldoftas, B. and Levine, D. I. (1999) "Flexibility versus efficiency? A case study of model changeovers in the Toyota production system", *Organization Science,* Vol 10, No 1, pp 43-68.

Bacharach, S.B. (1989), "Organizational theories: Some criteria for evaluation", *Academy of Management Review*, Vol 14, No 4, pp 496-515

Benner, M. J. and Tushman, M. L. (2003) "Exploitation, exploration and process management", *Academy of Management Review,* Vol 28, No 2, pp 238-256.

Birkinshaw, J., Hamel, G. and Mol, M. J. (2008) "Management innovation", *Academy of Management Review,* Vol 33, 825-845.

Burgelman, R. A. (1984) "Designs for corporate entrepreneurship in established firms", *California Management Review,* Vol 26, No 3, pp 154-166.

Christensen, C. M. and Bower, J. L. (1996) "Customer power, strategic investment, and the failure of leading firms", *Strategic Management Journal,* Vol 17, No 3, pp 197-218.

Christensen, C. M., Johnson, M. W. and Rigby, D. K. (2002) "Foundations for growth: How to identify and build disruptive new businesses", *MIT Sloan Management Review,* Vol 43, No 3, pp 22-31.

How to Design for Strategic Innovation? Appropriate Forms of Ambidexterity

Christensen, C. M. and Overdorf, M. (2000) "Meeting the challenge of disruptive change", *Harvard Business Review*, Vol 78, No 2, 67-76.

Cohen, W. M. and Levinthal, D. A. (1990) "Absorptive capacity: A new perspective on learning and innovation", *Administrative Science Quarterly*, Vol 35, No 1, pp 128-152.

Cooper, R. G. (2008) "The Stage-Gate (R) idea-to-launch process-update, what's new and nexgen systems", *Journal of Product Innovation Management*, Vol 25, No 3, pp 213-232.

D' Aveni, R. A. (1999) "Strategic supremacy through disruption and dominance", *Sloan Management Review*, Vol 40, No 3, pp 127-135.

Day, G.S. (2002), Managing the market learning process", *Journal of Business and Industrial Marketing*, Vol 17, No 4, pp 240-252.

Duncan, R. (1976) "The ambidextrous organization", In *The management of organization*, Vol. 1 (Eds, Killman, R. H., Pondy, L. R. and Sleven, D.) North-Holland, New York, pp. 167-188.

Eisenhardt, K. M. (1989) "Building theories from case study research", *Academy of Management Review*, Vol 14, No 4, pp 532-550.

Gibson, C. B. and Birkinshaw, J. (2004) "The antecedents, consequences and mediating role of organizational ambidexterity", *Academy of Management Journal*, Vol 47, No 2, pp 209-226.

Gopalakrishnan, S. and Damanpour, F. (1997) "A review of innovation research in economics, sociology and technology management", *Omega-International Journal of Management Science*, Vol 25, No 1, pp 15-28.

Govindarajan, V. and Kopalle, P. K. (2006) "Disruptiveness of innovations: Measurement and an assessment of reliability and validity", *Strategic Management Journal*, Vol 27, No 2, pp 189-199.

Govindarajan, V. and Trimble, C. (2005) "Organizational DNA for strategic innovation", *California Management Review*, Vol 47, No 3, pp 47-76.

Griffin, A., Price, R. L., Maloney, M. M., Vojak, B. A. and Sim, E. W. (2009) "Voices from the Field: How exceptional electronic industrial innovators innovate", *Journal of Product Innovation Management*, Vol 26, No 2, pp 222-240.

Gupta, A. K., Smith, K. G. and Shalley, C. E. (2006) "The interplay between exploration and exploitation", *Academy of Management Journal*, Vol 49, pp 693-706.

Hamel, G. and Getz, G. (2004), "Funding growth in an age of austerity", *Harvard Business Review*, 82, 7-8, 76-84

Hauser, J. R., Tellis, G. J. and Griffin, A. (2006) "Research on innovation: A review and agenda for Marketing Science ", *Marketing Science*, Vol 25, pp 687-717.

Hurley, R. F. and Hult, G. T. M. (1998) "Innovation, market orientation, and organizational learning: An integration and empirical examination", *Journal of Marketing*, Vol 62, No 3, 42-54.

Iansiti, M., McFarlan, F. W. and Westerman, G. (2003)"Leveraging the incumbent's advantage", *Sloan Management Review*, Vol 44, pp 58-+.

Jaworski, B., Kohli, A. K. and Sahay, A. (2000) "Market-driven versus market-driving", *Journal of the Academy of Marketing Science,* Vol 28, No 1, pp 45-54.

Johnson, J. L., Lee, R. P. W., Saini, A. and Grohmann, B. (2003)"Market-focused strategic flexibility: Conceptual advances and an integrative model", *Journal of the Academy of Marketing Science,* Vol 31, No 1, pp 74-89.

Kim, W. C. and Mauborgne, R. (1997) "Value innovation: The strategic logic of high growth", *Harvard Business Review,* 75, 103-112.

Kim, W. C. and Mauborgne, R. (1999) "Strategy, value innovation, and the knowledge economy", *Sloan Management Review,* Vol 40, pp 41-54.

Kim, W. C. and Mauborgne, R. (2000) "Knowing a winning business idea when you see one", *Harvard Business Review,* Vol 78, pp 129-138.

Kodama, M. (2003), "Strategic innovation in big business: Case studies of two Japanese companies", *Organization Studies*, Vol 24, No 2, pp 235-268.

Larsen, E., Markides, C. C. and Gary, S. (2002) *Imitation and the sustainability of competitive advantage,* Academy of Management, Denver.

Larsen, E., Markides, C. C. and Nattermann, P. M. (2003) "New entry, strategy conversion and the erosion of industry profitability", Strategic and International Management, London Business School, London, UK.

Lilien, G. L., Morrison, P. D., Searls, K., Sonnack, M. and von Hippel, E. (2002) "Performance assessment of the lead user idea-generation process for new product development", *Management Science*, Vol 48, pp 1042-1059.

Louis, M. R. and Sutton, R. I. (1991) "Switching cognitive gears: From habits of mind to active thinking", *Human Relations,* Vol 44, No 1, 55-76.

Lynn, G. S., Morone, J. G. and Paulson, A. S. (1996), "Marketing and discontinuous innovation: The probe and learn process", *California Management Review,* Vol 38, pp 8-37.

March, J. G. (1991) "Exploration and exploitation in organizational learning", *Organization Science,* Vol 2, pp 71-87.

Markides, C. (1997) "Strategic innovation", *Sloan Management Review,* Vol 38, No 3, pp 9-23.

Markides, C. (2006) "Disruptive innovation: In need of better theory", *Journal of Product Innovation Management,* Vol 23, No 1, pp 19-25.

Markides, C. and Charitou, C. D. (2004) "Competing with dual business models: A contingency approach", *Academy of Management Executive,* Vol 18, No 3, pp 22-36.

Matthyssens, P. and Vandenbempt, K. (2003) "Cognition-in-context: Reorienting research in business market strategy", *Journal of Business and Industrial Marketing,* Vol 18, No. 6/7, pp 595-606.

McDermott, C. M. and O'Connor, G. C. (2002) "Managing radical innovation: An overview of emergent strategy issues", *Journal of Product Innovation Management,* Vol 19, No 6, 424-438.

Miles, M. B. and Huberman, A. M. (1994) *Qualitative Data Analysis: An expanded sourcebook,* Sage Publications, Thousand Oaks, CA.

Mizik, N. and Jacobson, R. (2003) "Trading off between value creation and value appropriation", *Journal of Marketing,* Vol 67, No 1, 63-76.

Narver, J. C., Slater, S. F. and MacLachlan, D. L. (2004), "Responsive and proactive market orientation and new-product success", *Journal of Product Innovation Management,* Vol 21, No 5, 334-347.

O'Connor, G. C. (2008) "Major innovation as a dynamic capability: A systems approach", *Journal of Product Innovation Management,* Vol 25, No 4, 313-330.

O'Connor, G. C. and DeMartino, R. (2006), "Organizing for radical innovation: An exploratory study of the structural aspects of RI management systems in large established firms", *Journal of Product Innovation Management,* Vol 23, No 6, pp 475-497.

O'Reilly, C. A. and Tushman, M. L. (2004) "The ambidextrous organization", *Harvard Business Review,* Vol 82, No 4, 74-+.

Orton, J. D. (1997) "From inductive to iterated grounded theory: Zipping the gap between process theory and process data", *Scandinavian Journal of Management,* Vol 13, No 4, pp 419-438.

Porac, J. F., Thomas, H. and Baden-Fuller, C. (1989) "Competitive groups as cognitive communities: The case of Scottish knitwear manufacturers", *Journal of Management Studies,* Vol 26, No 4, pp 397-416.

Raisch, S. (2008) "Balanced structures: Designing organizations for profitable growth", *Long Range Planning,* Vol 41, No 5, pp 483-508.

Raisch, S. and Birkinshaw, J. (2008) "Organizational ambidexterity: Antecedents, outcomes and moderators", *Journal of Management,* Vol 34, No 3, pp 375-409.

Siggelkow, N. and Levinthal, D. A. (2003) "Temporarily divide to conquer: centralized, decentralized and reintegrated approaches to exploration and adaptation", *Organization Science,* Vol 14, No 6, pp 650-669.

Siguaw, J. A., Simpson, P. M. and Enz, C. A. (2006), "Conceptualizing innovation orientation: A framework for study and integration of innovation research", *Journal of Product Innovation Management,* Vol 23, No 6, pp 556-574.

Slater, S. F. and Narver, J. C. (1995) "Market orientation", *Journal of Marketing,* Vol 59, No 3, pp 63-74.

Slater, S. F. and Narver, J. C. (1998) "Customer-led and market-oriented: Let's not confuse the two", *Strategic Management Journal,* Vol 19, No 10, pp 1001-1006.

Spender, J. C. (1989), Industry Recipes: The nature and sources of managerial judgment, Basil Blackwell, Oxford.

Tashakkori, A. and Teddlie, C. (1998), *Mixed methodology: Combining qualitative and quantitative approaches,* Sage Publications, Thousands Oaks, CA.

Thomas, J. B., Clark, S. M. and Gioia, D. A. (1993), "Strategic sensemaking and organizational performance: Linkages among scanning, interpretation, action, and outcomes", *Academy of Management Journal,* Vol 36, No 2, pp 239-270.

Tripsas, M. and Gavetti, G. (2000), "Capabilities, cognition, and inertia: Evidence from digital imaging", *Strategic Management Journal*, Vol 21, No 10-11, pp 1147-1161

Tuominen, M., Rajala, A. and Möller, K. (2004), "Market-driving versus market-driven: Divergent roles of market orientation in business relationships", *Industrial Marketing Management*, Vol 33, No 3, pp 207-217.

Tushman, M. L. and O'Reilly, C. A. (1996) "Ambidextrous organizations: Managing evolutionary and revolutionary change", *California Management Review*, Vol 38, No 4, pp 8-30.

Volberda, H. W., Baden-Fuller, C. and van den Bosch, F. A. J. (2001) "Mastering strategic renewal: Mobilising renewal journeys in multi-unit firms", *Long Range Planning*, Vol 34, No 2, pp 159-178.

Von Hippel, E. (1988) *The sources of innovation*, Oxford University Press, New York.

Wolfe, R. A. (1994) "Organizational innovation: Review, critique and suggested research directions", *Journal of Management Studies*, Vol 31, No 3, pp 405-431.

Zollo, M. and Winter, S. G. (2002) "Deliberate learning and the evolution of dynamic capabilities", *Organization Science*, Vol 13, No 3, pp 339-351.

Competent to Innovate: An Approach to Personal Development to Improve Innovation Competency in SME's

John Howard

University of Central Lancashire, Preston, UK

First published in The Proceedings of ECIE 2010.

Editorial Commentary

As a continuation of earlier contributions which developed a maturity model framework for organizational change in assessing SME capability for innovation, this chapter explores learning at the individual level by addressing the changing competency requirements within SMEs that strive to develop their innovation capabilities.

Howard proposes a theoretical basis for an individual competency framework applied as part of an overall change management program. He identifies competency and development needs for individuals within the organization based on both their role and the innovation maturity of the organization, as determined by assessment against the maturity model. This model will be an interesting reference for competency management in order to develop innovation capability.

Abstract: This paper builds upon earlier work presented at ECIE in 2009 by Howard and Gillies. The Excellence in Innovation Framework (EiI) is a Maturity Model framework for organisational development and change management to enable

SMEs to assess their current capability in innovation in order to maximise the return on their investments and prioritise developments in organisational processes encapsulated within an on-line tool which is available to SMEs via the Internet. The paper describes the theoretical basis for a competency framework applied as part of an overall change management approach which identifies competency and development needs for individuals within the organisation based on both their role and the maturity of the organisation determined by assessment against the maturity model. Competency is assessed against performance levels which identify the individual's degree of skill or proficiency in relation to each identified competency item. Once development needs have been identified these can then be addressed in learning and development programmes tailored to the specific needs of the individual with re-assessment of competency producing an evaluation of the effectiveness of the learning.

Keywords: innovation, small medium-sized enterprise (SME), competency, maturity model, education, training

1. Setting and objectives

This paper builds upon this earlier work presented at ECIE in 2009 by Howard and Gillies and is part of the Knowledge to Innovate initiative (K2i) by the Northwest Regional Development Agency in response to the Regional Economic Strategy (2006) as part of the delivery against Transformational Action 12 to develop higher added value activity through innovation.

The EiI has been well received; however, it is apparent that change within business processes cannot bring about desired change in the absence of staff that are competent to work within those processes.

In earlier work the author, in collaboration, developed the EiI to

- Enable participating SMEs to assess their own capability for innovation
- Enable participating SMEs to identify their development priorities for improving the capability for innovation
- Enable the K2i programme to measure the capability of participating SMEs for innovation
- Enable the K2i programme to identify the development priorities for improving the capability for innovation of participating SMEs

The missing element from this approach is the staff working within the organisation. The Maturity Model focused on developing the organisation without explicitly considering the development of individuals within that organisation. The purpose of this paper is to advance the earlier work by addressing this "human" aspect of the organisation. The innovation process is not complete without connections being made at the level of skills, functions, technologies, commercial production, markets and other organisations (Mitra, 1999). Thus there is a need for a model and associated tools which will allow an SME to identify and address the knowledge, skills and attitudes of staff for innovation. This paper outlines such an approach.

2. Theoretical approach

2.1. Maturity modelling the organisation

The Excellence in Innovation framework is based upon principles first expressed in the Capability Maturity Model (Humphrey et al, 1989) and developed for other domains by the authors (eg Gillies and Howard, 2003, 2007). In November 1986, the US Government asked the Software Engineering Institute (SEI) to provide the federal government with a method for assessing the capability of their software contractors. A key part of the approach was an emphasis upon improvement, which contrasted sharply with other models of the time such as ISO9000, AQAP and BS5750.

The SEI capability maturity model (CMM) (Paulk et al, 1993a, 1993b; SEI, 1995) is defined as a five-level framework for how an organisation matures its software processes from ad hoc, chaotic processes to mature, disciplined software processes. The maturity model approach provides a paradigm within which domain-specific frameworks can be deployed to assist the capability development of small social and commercial enterprises.

The CMM has become an international standard in its field, and has spawned a whole range of "maturity models" mostly in software related areas, including the IT Service Capability Maturity Model, (Niessink and van Vliet, 1999) and other models aimed at project management and security. There is also an international standard, (ISO, 2008) which is based upon a similar maturity model approach.

2.2. The excellence in innovation framework – a maturity model approach

The EiI developed for Knowledge to Innovate (Knowledge to Innovate, 2009), assesses capability for innovation across seven dimensions: Collaboration, Environment, Finance, Knowledge, Senior Management, Risk, and Staff assessed against five maturity levels:

- 1. Commitment,
- 2. Putting a process in place,
- 3. Monitoring activity,
- 4. SMART goals and improving performance, and
- 5. Continuously improving performance.

2.3. Competency development – the performance ladder

There is some literature from the domain which sees learning as fundamentally an individual construct (e.g. Simon, 1991; Grant, 1996). However the general tendency in the business literature has been to consider learning as an organisational rather than an individual concept. This collective view of learning is compounded by the external locus assigned to "knowledge" in such an approach (Mitra, 2000). Lam (2004) acknowledges the individual role but argues that "both individuals and organisations are learning entities. All learning activities however, take place in a social context, and it is the nature and boundaries of the context that make a difference to learning outcomes". She thus stresses the collective and social element of learning as being the dominant factor.

Whilst this "organisational" perspective on learning may have some merits, it does tend to assume collective learning does not necessarily involve a learning process in the individuals who make up the social system of the organisation. It would seem self evident that, at its core, "learning" is a process that occurs within an individual; to move the focus of consideration away from this locus risks a fundamental misinterpretation of the processes involved. Kelly (1986) neatly summarised the various educational perspectives from the literature. Firstly; rationalist epistemology which offers certainty of knowledge and values leading to a simple transmission of the known from one generation to the next. The second is that of pragmatism; where education is essentially a process, with the focus on the "how" of learning as opposed to the "what" with an implied consensus on shared values. Finally Kelly suggests a third perspective, that of "New Directions", where many of the pragmatist assumptions are maintained but the pragmatic view of universally agreed values is dismissed as the imposition of the views of society's dominant groupings. Instead the focus is on the promotion of diversity of approaches. Given a need to develop innovation competency in SME's this final perspective would seem to offer a sound theoretical base with its emphasis on diversity.

It was probably McClelland (1973) who first introduced the concept of competence as a measure of performance as opposed to academic ability of knowledge retention. Since then there has been much work (e.g.; Boyatzis, 1982; Spencer and Spencer, 1993) which has attempted to use this concept as a theoretical underpinning for performance enhancement in the workplace. Simpson argues that "enthusiasm for this approach to competency has waned somewhat, due in part to the growing confusion about what the word actually means" (Simpson, 2002, p53). She also proposes that the concept of competency at the level of the individual is only weakly theorised and offers a competency framework for knowledge management incorporating 3 distinct dimensions; Knowing Why, Knowing How and Knowing Whom. Of these Simpson suggests that the knowing how dimension most easily fits into existing models of competency, however she does not refer to earlier work by Benner or Dreyfus and Dreyfus. These authors seem to provide not only a sound theoretical basis for, but definition of, the concept of competency and how it can be used to improve performance, in this case performance in Innovation. Benner's approach, building on that of Dreyfus and Dreyfus, incorporates all three dimensions described by Simpson.

Benner (1984) has outlined an adaptation of the Dreyfus model of skill acquisition (Dreyfus and Dreyfus, 1980) applied to nursing. Benner's model outlines a number of stages on the way to becoming a skilled practitioner based upon three aspects of overall performance. Firstly, a shift in paradigms from abstract rules to life experiences as the basis for behaviour. Secondly a change in perception of situations; from a collection of disparate equal parts to a complete entity in which some parts have more relevance or importance than others. Thirdly the move from "detached observer" to "involved performer". Within this framework there are five stages through which the student will pass on their way from a novice to becoming an expert. This scale of performance is adapted to provide the basis for the performance rating scale used in this model. See Table 1.

Using this model, skilled performance is neither a measure of outcome, nor behaviour, but an amalgam of both, and includes the way in which the individual processes information before acting, and the way in which they act to achieve a desired outcome.

There is an old saying relating to a room full of monkeys with a typewriter and their ability, given long enough to produce a copy of Shakespeare's "Hamlet". Whilst this may, or may not, be true it does highlight the idea that high quality outputs are not always the result of expert or skilled performance; rather they may be due to blind luck. Anyone, of reasonable intelligence and ability could, given a set of detailed enough instructions, undertake a range of highly complex activities achieving a desired output. This would not make them an expert. This is the level of performance described by Benner as Level 1 – Novice. Behaviour is dictated by adherence to context-free rules. The learner's actions are guided to provide achievement of a task. However performance is also constrained by these same rules as no account is taken of the relative importance of different components within the situation.

Table 1: Six levels of performance ladder (after Benner, 1984)

Level	Designation	Description
0	Unskilled /Not Relevant	The individual is unable to perform this skill even under instruction or the skill is not required in this role
1	Novice	The individual has little or no experience in this aspect. Able to perform only under close instruction or guidance.
2	Learner	The individual has some experience in this aspect and is able to perform with minimal day-to-day supervision but still requires regular instruction or guidance as new situations arise.
3	Competent	The individual performs in this aspect regularly and is able to work effectively, without supervision, on a day-to-day basis, but may need occasional instruction, guidance or support when confronted with unusual situations.
4	Proficient	Skilful in this aspect. The individual has a wealth of experience and functions with only managerial supervision. Is capable of demonstrating this aspect to others
5	Expert	Highly skilful in this aspect with several years experience. The individual has an intuitive grasp of the aspect and requires no supervision other than clinical governance. Acts as a mentor and innovator in this aspect.

As an Expert, the individual has passed beyond the need for what Benner refers to as "analytic principle" and is now able to adapt their actions. They no longer need to follow a list of instructions or cognitively refer each situation to the archive of past experience for comparison; their actions are now performed from an intuitive grasp of a situation.

However, a fundamental problem with the Benner model is that an individual's performance is not of necessity uniform in all aspects of their role. As the model is situational and experientially based, performance in disparate areas will necessarily be at different levels depending on theoretical knowledge and previous experience of the individual.

In order to address this, the performance rating scale used relates, not to the whole individual, as is the case in Benner's work, but to selected facets of performance expertise against each competency item.

The author has previously used this approach in a number of settings including the innovation of electronic patient records in a health care context (Department of Health: 1998, 2001). Traditionally the clinical record of a patient has consisted of pieces of paper with markings upon them with a coloured dye, hand writing. These marks being placed on the paper by those dealing with the patient as a record of their patients care.

This system required a series of skills from health care practitioners in order for them to access the information within this record or to add new information to it. These information management skills we commonly refer to as reading and writing. Whilst the clinical record was held as streaks of vegetable dye on dried out wood pulp these skills were appropriate, vital and all that was required, in addition to the obvious requirement for appropriate clinical or management skills. Once the clinical record is held as a series of digital bytes on a computer these clinicians and anyone else involved with the patients care, require new skills in order to record patient data, and most importantly leverage that data so as to make decisions on the highest quality of information.

3. Competent to innovate framework (C2I) – integrating competency with maturity

It is clear that, in order to reap practical management and administrative benefits, including competitive advantage, there is a need not only for the appropriate technology, systems and processes to be in place, but also to have staff available to operate these with an appropriate level of skill. (Storey et al, 2002)

Performance of skills can be viewed in a variety of ways. Bentley (1996) outlines two performance paradigms. Firstly, as a measurement of how well people are doing the things that they do; this is the competency based or process approach. Secondly, by disregarding the way in which things are done and simply looking at the outcomes. This is the output-focused or product approach to performance management. It is however possible to conceive of a model of performance which bridges these paradigms including what the individual brings to the situation, their skills with what they achieve, the output.

As new functionality is implemented, and new processes initiated, the skills requirements of various roles change. Thus. in order to maximize the value from investments a model for the identification of training needs is required which not only takes account of the persons role and the skills required for that role, but also the level of innovation maturity of the organization. For example, there is little value to be gained by training secretaries in e-mail when the organization uses carrier pigeons for its internal communications.

The competency items contained in the C_2I framework are derived from those identified by the implementation team as being those which are required if staff are to have the capacity to innovate. The items are derived from National Occupational Standards (NOS) which are widely used in both the public and private sector within the UK as a standard for the knowledge, skills and attitudes which may be required in given workplaces and industry sectors. For C_2I the NOSs were chosen from the UK Management Standards Centre's framework (MSC, 2008). These were then mapped to the Maturity Dimensions for EiI to ensure all aspects of Innovation were appropriately addressed. These are shown in Table 2.

A model combining the performance ladder outlined previously, with the innovation maturity of the organization was then developed. Firstly a two-dimensional competency matrix with competency items along one axis and professional roles on the other axis is constructed.

Each intersecting cell holding the performance required for that combination of role and competency item. This is the level of performance assigned to that particular combination of role and skill. (See figure 1).

Table 2: NOS derived competency items mapped to EiI maturity dimensions

Collaboration, MSC NOS A3 Personal networks MSC NOS D17 Collaborative relationships with other organisations MSC NOS F9 Market and customer understanding **Environment,** MSC NOS B2 Operational environment mapping Finance, MSC NOS E1 Budget Management MSC NOS E3 Financing **Knowledge,** MSC NOS E11 Communication MSC NOS E14 Team support MSC NOS D7 Provide learning opportunities for colleagues MSC NOS C1 Encourage innovation in your team MSC NOS C2 Encourage innovation in your area of responsibility MSC NOS E12 Manage knowledge in your area of responsibility MSC NOS E13 Promote knowledge management in your organisation **Senior Management,** MSC NOS B8 Legal, regulatory, ethical and social requirements MSC NOS B9 Organisational culture development MSC NOS C3 Promoting innovation MSC NOS C4 Change Leadership **Risk,** MSC NOS B10 Risk management **Staff** MSC NOS C6 Change implementation MSC NOS D8 Staff support MSC NOS D9 Team building and management MSC NOS D10 Conflict management MSC NOS F16 Product / Marketing

Level 0	Competency				
Role	A	B	C	D	E
Managing Director	2	4	3	4	3
Labourer	2	2	3	0	1
Secretary	0	3	3	3	2
Technician	4	4	1	1	1
Stores	2	1	2	2	1
Finance	3	2	3	3	3

Figure 1: Horizontal cross section of C_2I matrix

When innovation maturity is added, as a third axis the single layer matrix becomes a six-layer matrix with each layer representing the maturity of the organisation as determined by an Excellence in Innovation Maturity audit. Figure 2 shows an overview of this matrix; whilst figure 3 shows a cross section, identifying the different performance levels required by a Managing Director in a variety of competencies at increasing levels of EiI maturity.

As the SME matures its innovation capability, by utilising the EiI model, the skill levels required by each different role within the company will vary. Normally the performance level required will increase. However there may well be some instances where, for example role diversification with new staff being taken on, may actually lead to a decrease in the required performance level for an individual as that functions passes to a new member of staff. The model also allows for the addition and removal of competency items from the list of those currently required by the SME. In this way the competencies and performance required by each role may be constantly updated and revised in the light of change whilst retaining the history of each individual's competency development and the history of the role specifications required on a given date.

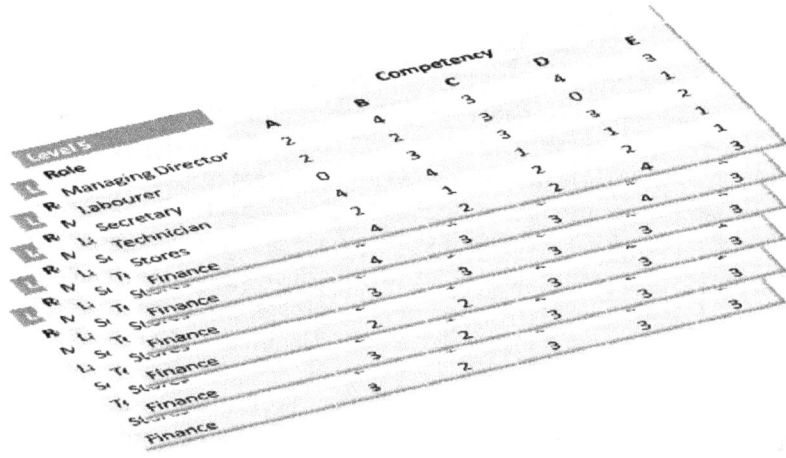

Figure 2: Overview of C$_2$I matrix

Role Finance	Competency				
	A	B	C	D	E
Level 0	1	1	1	1	1
Level 1	2	1	1	2	2
Level 2	3	2	2	3	3
Level 3	4	2	3	3	3
Level 4	4	2	3	4	3
Level 5	4	2	2	4	3

Figure 3: Vertical cross section of C$_2$I matrix

In order to understand the model fully it is probably simplest to "walk-through" the process of identifying an individual's training and development needs and then explore how these can be met.

Firstly the SME undertakes an EiI audit of its Innovation maturity. The SME could manage this process by an internal audit however experience has shown that the approach is most effective when used by an external facilitator / mentor. Once the EiI level is assessed the software selects the performance level required by their role, at the current level of maturity for their company, from the C_2I matrix. Effectively the appropriate row is "pulled" from the appropriate layer of the C_2I matrix so that there is a row of performance levels relating to the individuals role in a number of skills. These act as a benchmark against which performance can be assessed. Individual workers are then assessed against these benchmarks to identify any gap in performance and thus identify training needs. This assessment may be performed in a number of ways, including self-assessment. The preferred methodology, in most settings, is through a consultation process where the manager and the individual gather information from colleagues before jointly completing the assessment. This provides a triangulation of performance measures from the individual, the manager and objective output data. The results then form the basis of the individual personal development plan. They are also utilized by both the SME and the Development Agency to manage training for the coming year.

Whilst the author's preferred scale is that derived from Benner and outlined in table 1, the fundamental matrix model and software is equally applicable using any other performance-rating scale, such as that described by Boydell and Leary (1996) that uses a scale of 7 modes to describe an individual's performance level.

4. Education to improve innovation competency

Once a gap between the required skill level and the benchmark standard for an individual has been identified, a means must be found of addressing this need in order to help the individual move to the performance level required. It should be remembered that we are not simply concerned with outcome, nor with the behaviour undertaken, but the expertise with which the skill is performed to produce the desired outcome. Thus both process and product methodologies are embraced.

Central to the model are the links to educational resources associated with each competency item and the various skill levels. The resources selected being dependent on the individuals' current skill level, the level required of them, the competency in question and the maturity of the SME as assessed by the EiI level.

Once an individual's assessment has been undertaken the software will search for the educational resources associated with that competency and level to allow a bespoke educational programme to be designed based upon the identified needs. These resources may vary from a peer in another company within a cluster (European Cluster Alliance, 2009), through a library reference, a URL on the internet containing appropriate information or activities, to study days or full educational programmes. This is possible because every individual's needs are so highly specified by the system. Education and training provided can thus be focused on real needs. In addition the effectiveness of development activity can be measured by a re-assessment of the individual thus producing a pre and post training comparison to assess training effectiveness.

Following on from the educational or training input a reassessment of the competency in the workplace is repeated. This allows for a Kirkpatrick (1959) level 3 evaluation of the application of the learning in terms of work place performance. Indeed if the educational input is mapped to organisational goals and objectives this data may well be useful in any Kirkpatrick level 4 evaluation; essentially the effect on the "bottom line" for the SME.

5. Application

As part of the academic evaluation of the initial work, the development and implementation of the Excellence in Innovation Framework, a small Delphi type study was conducted with the implementation team. From this a number of issues were raised which fed into the development of the competency framework described here. A number of points were identified which are presented as advantages and disadvantages of the approach below.

5.1. Advantages

1. In isolation the EiI maturity model approach addresses only one of the two key components of the organisations capacity to innovate: the structural, environmental and process aspects of the organisational entity. There is also a fundamental need for an approach to address the human elements. This approach provides a clear and simple framework for staff development which links the required knowledge, skills and attitudes of the level of maturity of the systems, policies and procedures of the SME within which they work.

2. The approach allows business supporting entities such as Clusters and Regional Development Agencies to identify needs across a geographical area and analyse those needs to enable the development of appropriate education and training which would not be possible if the SME was left to its own devices.

3. The clearest message to come from the implementation team was that in those SMEs where the staffs opinions and ideas were really valued the culture of innovation was almost palpable and was followed by clear improvements in the maturity of the organisation within the EiI model. The consideration of individual learning and development as being of equal value to the organisation as more structural considerations such as policies and procedures means that the staff are not only allowed but encouraged to question everything they do producing a synergy of cultural change enhancing innovation.

5.2. Disadvantages

1. Additional complexity. One of the key barriers to the implementation of this approach which was highlighted by the implementation team was the size of the SME. In those SME's where the workforce was under 10-15 staff it was observed that the use of the underlying maturity model for innovation was of limited value in new and micro businesses where the business felt there was simply insufficient time for such activity with all energies being focused on survival.

2. Clearly the use of this approach requires an investment in time from the SME. Where the SME is still at an early stage in its growth it may well be that there is simply not the time and energy available to undertake such strategic activities no matter how important they are for long term growth. This additional workload can be minimised by the use of mentors and facilitators working for support agencies such as the Northwest Development Agency.

3. Finally, the implementation team found examples of SMEs which this approach would simply not work for. These could be characterised as those where the entrepreneur adopted an alpha male approach and would not allow staff the freedom to question any decisions. In such SMEs it was felt that neither the Excellence in Innovation or a Competent to Innovate framework would be of value but such organisations would seem to have a limited life expectancy anyway.

6. Summary

This model offers a simple, yet effective, tool for managing the changing competency requirements of staff within an SME attempting to develop its innovation capability. The synergistic effect offered by the bringing together of process maturity (EiI) and competency (C_2I) is potentially huge. By using the maturity of the SME in a given change process, both current and prospective, as the basis for development need assessment of staff, education and training provision is focused on the areas actually needed at any given time in order to support the change process. By individual assessment against the competency framework, customized learning plans can be provided for each worker which gives them not simply a list of the things they fail to do adequately. Instead it points them to the resources they require in order to help them develop their skills to those that are now required of them. The role of a supporting network such as that offered by the cluster concept and the development agency system in the UK regions.

7. Acknowledgements

The author gratefully acknowledges the input of K2i staff, notably Bea Acton and Geoff Birkett and the support of the North West Development Agency for the original work.

References

Benner, P. (1984). From Novice to Expert, Excellence and Power in Clinical Nursing Practice. Addison-Wesley, Manlo Park CA.

Bentley, T.J. (1996). Bridging the Performance Gap, Gower, Aldershot

Boyatzis, R. (1982). The competent manager, John Wiley and sons. New York

Boydell, T. Leary, M. (1996). Identifying Training Needs, Institute of Personnel and Development, London

Department of Health. (1998). Information for Health. HMSO, London. Available on the web at
http://www.dh.gov.uk/en/Publicationsandstatistics/Publications/PublicationsPolicyAndGuidance/DH_4002944

Department of Health. (2001). Building the Information Core – implementing the NHS Plan HMSO, London. Available on the web at
http://www.dh.gov.uk/en/Publicationsandstatistics/Publications/PublicationsPolicyAndGuidance/DH_4005249

Dreyfus, S.E. Dreyfus, H.L. (1980). A five stage model of the mental activities involved in directed skill acquisition. Unpublished report supported by AFSC, USAF contract F49620-79-C-0063, University of California at Berkeley

European Cluster Alliance. (2009). May 2009 Report: "Improving the Cluster infrastructure through policy actions"

Gillies, A.C. Howard, J. (2003). Managing change in process and people: combining a maturity model with a competency-based approach TQM & Business Excellence, vol. 14, no. 7, September, 797–805

Gillies, A.C. Howard, J. (2007) Modelling the way that dentists use information: an audit tool for capability and competency, British Dental Journal, November 2007, Volume 203 No 9, pp529 – 533.

Grant, R.M. (1996). 'Toward a Knowledge-Based Theory of the Firm'. Strategic Management Journal, 17: 109-122.

Humphrey, W. Sweet, W.L. Edwards, R.K. LaCroix, G.R. Owens, M.F. Schulz, H.P. (1987). Method for Assessing the Software Engineering Capability of Contractors, A , Carnegie Mellon Software Engineering Institute Technical Report CMU/SEI-87-TR-023 ADA187230, available on the web at http://www.sei.cmu.edu/publications/documents/87.reports/87.tr.023.html

ISO (2008) ISO/IEC TR 15504-7:2008 Information technology -- Process assessment -- Part 7: Assessment of organizational maturity. International Organization for Standardization, Geneva.

Kelly, A.V. (1986). Knowledge and curriculum planning, Harper and Row, London.

Kirkpatrick, D.L., (1959). Techniques for evaluating training programs. Journal of ASTD 11, pp. 1–13.

Knowledge to Innovate. (2009). The Knowledge to Innovate programme, funded by the Northwest Regional Development Agency, details available on the web at www.k2i.org.uk

Lam, A. (2004). Organisational Innovation. Working paper 1, BRESE, Brunel University.

M.S.C. (2008). Management Standards Centre, National Occupational Standards for Management and Leadership 2008. Available on the web at: www.management-standards.org/content_1.aspx?id=10:5406&id=10:1917

Mitra, J. (1999). "Managing externalities: integrating technological and organisational change for innovation". Keynote paper presented at the 9th international technology forum. 4th-8th October, University of Minnesota, Minneapolis, MN

Mitra, J. (2000). Making connections: innovation and collective learning in small business, Education and Training, Vol 42 No 4/5 pp 228-236

McClelland, DC. (1973). Testing for competence rather than intelligence. American Psychologist, 28:1-14

Niessink, F. van Vliet, H. (1999). The Vrije Universiteit IT Service Capability Maturity Model.

Vrije Universiteit Amsterdam. Available on the web at
  http://www.serc.nl/people/niessink/publications/TR99.Niessink.pdf
North West Regional Development Agency, (2006). Northwest Regional Economic
  Strategy 2006. Available on the web at
  http://www.nwda.co.uk/publications/strategy/regional-economic-strategy-
  200.aspx
Paulk, M. C. Curtis, B. Chrissis, M.B. Weber, C.V. (1993). "Capability Maturity Model
  for Software, Version 1.1", Carnegie Mellon Software Engineering Institute
  Technical Report CMU/SEI-93-TR-24, DTIC Number ADA263403, Available on
  the web at http://www.sei.cmu.edu/reports/93tr024.pdf
Paulk, M.C. Charles, V. Garcia, S.M. Chrissis, M.B. Bush, M.W. (1993). "Key Practices
  of the Capability Maturity Model, Version 1.1", Carnegie Mellon Software Engi-
  neering Institute Technical Report CMU/SEI-93-TR-25, DTIC Number
  ADA263432, Available on the web at
  http://www.sei.cmu.edu/publications/documents/93.reports/93.tr.024.html
Simon, H.A. (1991). 'Bounded Rationality and Organizational Learning'. Organiza-
  tion Science, 2: 125-134.
Simpson, B. (2002). The knowledge needs of innovating organisations, Singapore
  Management Review, Vol 24, No 3 pp51-60
Software Engineering Institute, (1995). Carnegie Mellon University The Capability
  Maturity Model®: Guidelines for Improving the Software Process Addison-
  Wesley ISBN: 0-201-54664-7,
Spencer, I.M. Spencer, S.M. (1993). Competence at work: models for superior per-
  formance. John Wiley and Sons, New York
Storey, L. Howard, J. Gillies, A. (2002) Competency in Healthcare, Radcliffe, Oxford

Why Intellectual Capital Management Accreditation is a Tool for Organizational Development?

Florinda Matos[1, 2], Albino Lopes[1], Susana Rodrigues[2] and Nuno Matos[3]

[1]ISCTE - Lisbon University Institute, Portugal
[2]ESTG - Polytechnic Institute of Leiria, Portugal
[3]PMEConsult, Portugal

First published in the Electronic Journal of Knowledge Management (www.ejkm.com) Vol 8, issue 2, 2010.

Editorial Commentary

Intellectual Capital is dais to be the firm's capacity to transform knowledge and other intangible assets into wealth-building resources (Edvinsson, 2002). It is possible to link intellectual capital (as a source of advantage) and innovation (as a source of competitive advantage) since some consider that firms with superior intellectual resources outperform competitors in exploring, deploying, combining and configuring resources and capacities in distinctive ways to create customer value (Spender & Marr, 2004; Santos-Rodrigues et. al., 2010).

This chapter, authored by Matos, Lopes, Rodriguez and Matos, deals with Intellectual Capital Management Accreditation as a significant device for Organization Development. It discusses to what extent the accreditation process and the associated methodology increases the corporate innovation dynamic. Accreditation is presented as a rigorous process of valuation which enhances internal capabilities, good practices, knowledge sharing and innovativeness. The approach reflects a logical and original process

References

Edvinsson, L. (2002). What is IC? www.unic.net. Retrieved 2008

Santos-Rodrigues, Helena; Dorrego, Pedro Figueroa; Jardon, Carlos Fernandez. The Influence Of Human Capital On The Innovativeness Of Firms. International Business & Economics Research Journal, Sep 2010, Vol. 9 Issue 9, p53-63

Spender, J.-C, & Marr, B. (2005). Knowledge-based perspective on Intellectual Capital. In B. Marr (Ed.), Perspectives on intellectual capital (pp. 183-195). Oxford: Elsevier Butterworth-Heinemann publications

Abstract: In March 2000, the European Council held an extraordinary meeting to agree a new strategic goal for the European Union in order to strengthen a knowledge-based economy. The Council has a strategy - the Lisbon Strategy - aiming in the next 10 years to make the EU the most competitive and dynamic knowledge-based economy in the world. Intellectual capital has become a key element of the knowledge economy. Its management is therefore a factor influencing the competitive advantage of companies, regions and even countries. The purpose of this paper is to discuss the importance of intellectual capital management accreditation as a factor in the organizational development of companies, especially small and medium-sized enterprises (SMEs). The methodology ICMA - Intellectual Capital Management Accreditation (Matos and Lopes, 2009) will be discussed here, as well as the effect of this methodology on SMEs' innovation processes. It is considered that intellectual capital management accreditation may be a relevant process in the consolidation of an innovative dynamic, which will contribute to the continuous creation of competitive advantages. There are various intellectual capital valuation methodologies, but the research about the effect of the certification and accreditation is still very limited so it is necessary to get more results. However, the methodological research that supported the ICMA system points to the fact of accreditation procedures favouring better management of intellectual capital, thus contributing significantly to improving the organizational performance of accredited companies. This paper also aims to contribute to the international recognition of the importance of the audit of intellectual capital.

Keywords: intellectual capital management, ICMA, accreditation

1. Introduction

The environment in which businesses operate has changed substantially. The most valuable and productive assets do not appear on the Balance sheet and the traditional tools do not allow us to know what influence they have on business performance. The financial indicators appear not to be sufficient because they do not tell us whether we are increasing our competitive advantages.

Empirical studies, conducted by Matos and Lopes (Matos and Lopes, 2008) showed that the real competitive advantage results in, increasingly, the management of intangible assets. ICMA Methodology - Intellectual Capital Management Accreditation (Matos and Lopes, 2009) is designed precisely to fill this gap in assessing the management of intangible assets.

The various investigations carried out in Portuguese SMEs demonstrate that the high innovative potential of some SMEs may be recognized and enhanced through ICMA, which is thus a tool capable of enhancing the SMEs' competitiveness. In fact, business innovation is essentially incremental and routines are very important in supporting this type of innovation. The accreditation function is to monitor and to guide these routines. Intellectual capital management accreditation can, therefore, be very important in reducing the variance and in consolidating the innovation process.

The purpose of this paper is to make a critical exploration of intellectual capital management accreditation as a factor inducing dynamic innovation in SMEs. Over the next few sections of this paper we will discuss accreditation and demonstrate its importance for the consolidation of intellectual capital management as an organizational driver.

2. Literature review: Intellectual capital

Since we will analyze the process of accreditation of intellectual capital management, it is important to understand the concepts of intellectual capital, through the interpretation of some academics who have studied the issue. There are various definitions of intellectual capital and the concept continues to have a degree of subjectivity.

Why Intellectual Capital Management Accreditation is a Tool for Organizational Development?

Different words have been used to describe the concept of intellectual capital: intangibles, knowledge-based, and non-financial assets are some examples.

Sveiby, (1997), developed a measurement methodology, "The Intangible Asset Monitor", by dividing the intangible assets into three groups: individual skills, internal structure and external structure.

This author considers that the skills of the employees of a company are an intangible asset which, together with the other intangible assets, is added to the tangible assets, becoming the full assets of the organization.

Edvinson and Malone (1997), proposed a model, "Skandia Navigator" which divides intellectual capital into two categories: human capital and structural capital. Human capital is, according to these authors, the capital of the human resources in the company, consisting of its skills, the accumulated value of its practices, its creativity, its relationship capacity, its values, etc.. Part of this capital is also the culture and the organizational values of the company. In the opinion of the authors, it is this capital which is the source of innovation and renovation.

Structural capital, on the other hand, is understood as the value left in the company by the human resources when they go home, for instance, the database, the manuals, the list of clients, etc.. This capital can still be divided into organizational capital and client capital. And, in turn, organizational capital is divided into process capital, innovation capital and client capital.

Thus, according to this vision, intellectual capital is the sum of structural capital and human capital, this being the basic capacity for the creation of high quality value.

To Brooking (1996), the concept of intellectual capital arises from the association of different intangible assets, split into four categories: market assets, human assets, ownership of intellectual assets and assets of substructures.

Roos (1997) has a similar concept of intellectual capital to that of Edvisson and Malone (1997), but he considers intellectual capital as a result of the interaction between human capital, infrastructure capital and relationship capital.

Andriessen (2005) defines intellectual capital as "all intangible resources that are available to an organization, that give a relative advantage, and which in combination are able to produce future benefits."

We define intellectual capital as "an intangible element, resulting from the sum of knowledge of each individual in an organization arising from the wealth of people in the organization, their level of education, their experience, their information and their willingness to develop the acquisition of knowledge - i.e. individual talent"

Intellectual capital is divided into individual capital, team capital, processes capital and clients capital (See Intellectual Capital Model – ICM, Matos and Lopes, 2009)

Despite these differences in the classification of intellectual capital, it appears that these authors present unanimously the following points:

- Intellectual capital is an intangible asset that needs to be managed.
- Management of intellectual capital can create value in the organization.
- Management of intellectual capital can generate competitive advantages.
- Human capital, Clients capital and Processes capital are the main components of intellectual capital.

It is assumed as unquestionable the importance of intellectual capital as a factor inducing business development. By creating systems of certification and accreditation, we are looking for tools to help entrepreneurs in managing this valuable resource.

3. What is accreditation?

The concept of accreditation is not unique and often we find some confusion between certification and accreditation.

In Portugal, Law No. 125/2004 of 2004-05-31, defines accreditation as "the procedure by which the national accreditation body recognizes formally that an entity is technically competent to perform a specific function specifically, in accordance with international standards, European or national based, in addition, the guidelines issued by international accreditation bodies to which Portugal is party."

However, we find differences between the accreditation of intellectual capital management and accreditation systems of higher education (see Matos, 2008). In Portugal, Law No. 1 / 2003-01-6, defines the concept of academic accreditation as a "verification of the fulfillment of the requirements for the establishment and registration of courses"

Academic accreditation corresponds therefore to an official recognition of an institution or course, assuming an assessment based on pre-established standards, which serve as reference levels and for determining if the institution falls within those parameters, facilitating the recognition of diplomas, or degrees, by the legislator.

The significance of accreditation is therefore usually associated with official recognition and quality assurance, that is, to general acceptance. We can thus say that the purpose of accreditation is to ensure certain standards of quality. Based on previous concepts, the accreditation of intellectual capital management is a public statement that the company meets a set of established criteria for accreditation by the Accrediting Body.

The significance of accreditation is, therefore, usually associated with an operation of technical validation and recognition of the overall capacity of the entity, making it a member of a sort of "club" where they create the conditions for the dominance of best practices that make the accredited entity continuously seek alignment with the best performance.

4. What is ICMA methodology?

ICMA Methodology - Intellectual Capital Management Accreditation (Matos and Lopes, 2009) presents itself as a tool for the development of the accreditation process. This methodology has been developed progressively from a variety of research studies and aims to become the highest standard of recognition of intellectual capital management.

Companies with this accreditation have a commitment to quality and continuous improvement of the management of their intellectual capital. ICMA is a process that looks at the overall performance of the company and is designed to promote the skills of intellectual capital management with a view to innovation and sustainable competitiveness.

The methodology that supports ICMA is the result of theoretical research and several empirical research studies conducted over the past four years (see Lopes and Matos, 2005; Lopes and Matos, 2006; Matos and Lopes, 2008; Matos, 2008; Matos and Lopes, 2009).

The accreditation is based on the evaluation of a set of parameters - ICMA indicators. These indicators, allow us to evaluate the management of intellectual capital of companies, checking that there is evidence of the presence of indicators related to the dimensions of intellectual capital, and whether they are valued and managed.

The ICMA criteria are based on the ICM - Intellectual Capital Model which consists of 4 Quadrants specified by twenty five parameters (Matos and Lopes, 2009). To achieve ICMA, companies have to demonstrate that they meet the ICMA parameters in 4 areas: Individual Capital, Team Capital; Processes Capital and Clients Capital.

The Quadrant Individual Capital, Team Capital and Processes Capital are related to the company's internal environment, the Quadrant Clients Capital is related to the external environment.

In ICM, (See Figure 1) Individual Capital is called the Tacit Knowledge / Human Capital Quadrant. It is the knowledge inherent to the individual himself, and containing the real source of value, talents and the skills to generate innovation. Here, one has included the theoretical and practical knowledge of the individuals and the capacities of different types, such as artistic, sporting or technical.

Team Capital is the Human Capital / Explicit Knowledge Quadrant. The team shares the explicit knowledge. In this area, knowledge applies to the individual in the form of facts, concepts or tools.

When Explicit Knowledge is associated with Structural Capital, we are in the presence of applied experience, as the whole organization is the holder of formalized knowledge, able to be passed on, this is the Processes Capital. This Quadrant represents the ensemble of shared knowledge, summed up by experts (scientific community), recognized as the most advanced form of knowledge. This type of knowledge covers, among other dimensions, the organizational routines or the organizational memory. Organizational memory represents the register of an organization, represented by a set of documents and artefacts. Its goal is to expand and amplify knowledge through its acquisition, organization, dissemination, usage and re-finement. Organizational memory can be a way of registering tacit knowl-edge, making it explicit, so that through business processes it becomes part of the patrimony of the company, to be shared and recreated.

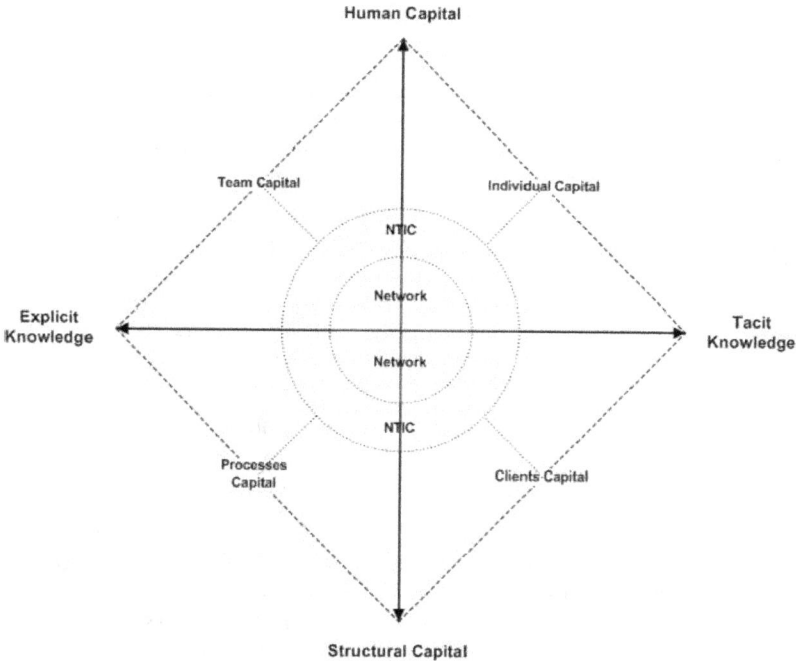

Figure 1: ICM - intellectual capital model

Clients Capital is the result of the interaction Structural Capital / Tacit Knowledge. This typology represents the organizational knowledge in its practical form and is already incorporated into the tacit experiences for-malized in the team. This knowledge, although hidden, becomes accessible through interaction, and it is the principal characteristic of the perform-ance of highly specialized teams.

In the Model presented, the Network and NTIC are essential in the rela-tionship between the 4 Quadrants.

Thus, the companies that put the NTIC at the service of human resources have a great advantage, because they can reduce the administrative difficulties in solving simple problems, increase the quality of services and promote continuous improvement and personal growth.

The approach to the concept of Network is not a new concept. The network, as a social concept, is the genesis of the social constructs of individuals. More recent is the approach to the concept of network system as a factor in the acquisition of knowledge and innovative action.

In conclusion, the NTIC are crucial to be effective Networks.

It should be noted that the ICM is a dynamic model and therefore is not a completely stabilized model. Thus, as it is applied to more companies, there may be further adjustments. Indeed, this is one of the advantages of the model: its interactive dynamism, which has proved very good in turbulent business contexts.

ICM parameters are:

Individual Capital Quadrant

- Use of NTIC: New technologies are an essential tool for company's organizational development. The purpose of this parameter is to demonstrate your domain for all employees.
- Networks: The networks, supported by new technology, are essential for the development of a networking culture. The purpose of this parameter is to prove the existence of an internal network with knowledge and talents that the company can use.
- Training / Qualification: Training / qualification are seen as the empowerment of individual employees. The purpose of this parameter is to examine how the company encourages the acquisition of knowledge and develop the talents of each of its employees.
- Valuation of Know – How: All employees of an organization have an inexhaustible stock of knowledge. However, often companies do not value and do not encourage these skills. Thus, the purpose of this parameter is to see how the company rewards and encourages the development and availability of knowledge and individual skills of their employees.
- Investment in Innovation and Development (ID): Innovation is a source of competitive advantage of companies. The purpose of

this parameter is to check whether the investment in ID, conducted by the company, aims to simplify processes or innovation.

Team Capital Quadrant

- Use of NTIC: New technologies should be used as a management tool, integrated in a networking culture. The objective of this parameter is to see, how the new technologies are used in building a team culture.
- Networks: he networks are forums for sharing knowledge and enable the dissemination of good practices. The purpose of this parameter is to demonstrate that the company promotes the existence of a network culture, where the teams interactive control, discuss and improve the procedures quality in order to satisfy the clients.
- Training / Qualification; Training / qualification should be understood as an instrument that enables the exchange of synergies between the organization employees. The company must have a policy of training and qualification perfectly synchronized with the team culture. The aim of this policy is to transform the group into cohesive, highly motivated and productive teams. The purpose of this parameter is verifying the existence of this policy of training and qualification.
- Team Work: The work must be organized into teams whose size will be most appropriate to the needs of the company. This parameter must show a teamwork culture.

Processes Capital Quadrant

- Use of NTIC: The company should use the new technologies as an administration tool, maximizing the use of these technologies in their organizational performance. New technologies are very important in the register of organizational knowledge and the operationalization of the whole process. The purpose of this parameter is to demonstrate how new technologies promote the improvement of procedures.

- Networks: This parameter enables us to evaluate how the company uses the "networks", articulated with the NTIC, to improve the processes and create interactivity between different stakeholders.
- Processes Systematization: The purpose of this parameter is to confirm process systematization and if it allows the formalization and transfer of knowledge among stakeholders.
- Existence of Certification: Companies should be granted certification, including the ISO 9001 certification. This parameter should confirm the existence of certification.
- Registration of Organizational Knowledge: Organizational knowledge must be registered. These records should be computerized in order to be protected and easily be shared. This parameter must verify the operability of the record of organizational knowledge.
- Partnerships: his parameter must verify the existence of a network of partnerships with various stakeholders.
- Investment in Innovation and Development (ID): The company must demonstrate how innovation and development enable connection and simplification of procedures. The parameter should demonstrate such evidence.
- The Brands Creation and Management: The purpose of this parameter is to demonstrate how the company's strategy relies on a process of creating and managing brands, which enables improved reliability of products or services and organizational differentiation.
- Complaints System: The company should have a formal system for registering complaints that serves its relationship with customers. The purpose of this parameter is to demonstrate the proper functioning of this complaints system.
- The existence of Awards: The awards are understood as the recognition of the process / customer relationship, resulting from the interaction of explicit knowledge with the structural capital. The purpose of this parameter is to check whether the company was awarded as a result of this recognition.

Clients Capital Quadrant

- Use of NTIC: This parameter must verify the functionality of the use of NTIC in improving the quality of service and interaction with customers.
- Networks: The networks should be part of an "act of collective intelligence" in which the expertise of each employee of the company is put at the service of customer satisfaction. The parameter must verify the existence of these networks, as part of the company's culture.
- Market Audits: Systematic market audits should enable the company to view the market where it will identify opportunities and threats. The purpose of the parameter is to check if the company performs these audits as part of their strategy.
- Management of the Clients' Satisfaction: The analysis of clients' satisfaction should be part of the company's organizational routines. Reports should be obtained, allowing the management of the company's relationship with clients. This parameter should check how the company manages its relationship with clients.
- Complaints System: This parameter must demonstrate that the complaints system, in addition to being part of a process, is an intrinsic element in the company culture.
- New Markets: The purpose of this parameter is to check if the company has a market strategy, in which the internationalization is one of the goals. The strategies of the market must be accompanied by strategies for innovation of products and services for new markets.

The various studies that have been made allow us to conclude that the ICMA process empowers intellectual capital, converting it into an innovation mechanism.

5. Is accreditation a tool for organizational development?

We live in a knowledge society which has seen the transition from a product economy to a service economy. Not only are we concerned with the process of creating valued products but also how the customer uses those products. This requires a broad collection of data across the value chain. This data is valuable because it can become information, which is transformed into knowledge that can produce innovation.

This innovation can be created from two types of resources: talented individuals that create disruptive innovation at the level of process, product or market, and incremental innovation that is supported by all workers, in the value chain, that are knowledge workers.

However, these knowledge workers are able to introduce micro-innovations that could continuously improve the value chain. So, if business innovation is essentially incremental, depending on system management, accreditation based on the audit of the management of intellectual capital gains importance.

Accreditation has the ability to put in the value chain of organizations, a surveillance system for each of its elements. Accreditation facilitates the monitoring of organizational routines. These routines, when based on the intellectual capital management of the whole team, generate creativity.

Accreditation, by imposing rules on intellectual capital management, requires organization and discipline that can generate dynamic innovation.

The university is the support of this new paradigm of the knowledge economy since this institution has the task of transforming knowledge into innovative and marketable products. Thus, we find the model of the university the inspiration for a model for the company that is complemented by the principles of quality assurance.

In a context of economic globalization, the institutions of higher education are required to review their development strategies and integrate their activities on an international plane.

The policies for developing and even funding higher education are currently based on models of development that are based on accreditation systems. It is believed that these models are a source of innovation, quality and competitiveness.

There is evidence that, in some higher education institutions, the academic quality of products is currently limited by several aspects, including: the growing number of institutions of higher education, internationalization of higher education systems, competition between educational institutions at the national and international level, the Bologna Process and the attempt to harmonize programs. The adoption of accreditation systems has become an essential instrument to promote and guarantee quality.

Why Intellectual Capital Management Accreditation is a Tool for Organizational Development?

At a time, when the labor market has become global and companies recruit their employees in various countries, institutions of higher education live in a dynamic environment, accompanied by increased mobility of teachers, and students who require a quick adjustment and anticipation of emerging trends. Particularly affected by this phenomenon, management schools and universities are placed in competition at both the national and international level, where they have to compete with the best foreign institutions.

In the field of management sciences, several international tables are used to measure the performance of education institutions, among the most representative include the following: AACSB (Association to Advance Collegiate Schools of Business), AMBA (Association for Masters of Business Administration) and EQUIS (European Quality Improvement System). Among these accreditation systems, EQUIS is the most international with more than 100 accredited schools, throughout the world.

EQUIS accreditation is an international system of quality evaluation, implementation and accreditation for Higher Education Institutions that have courses in the area of management and business administration. EQUIS is based on a group of principles which facilitate benchmarking, mutual learning and the dissemination of good practices. It is a multicultural and global system but of European inspiration.

Considering the analogy that we have been making between the accreditation systems of higher education and accreditation of intellectual capital management, we wanted to know the effect of accreditation in the innovative performance of higher education institutions. Thus, we analyzed the effect of this accreditation system in the performance of two Portuguese universities.

6. EQUIS empirical research

6.1. Methodology

This empirical research consists of 2 interviews with the managers responsible for the accreditation system EQUIS in two Portuguese Universities - Catholic University and New University of Lisbon, both of which have the accreditation system.

The first is a private higher education institution and the second a public higher education institution.

The semi-structured interviews are each about 1.5 hours and were recorded. The interviews were transcribed and their contents were analysed.

The processing and analyses of the information had the following stages:

- Creation of a content analysis matrix;
- Processing of statistical information;
- Application of the Support Model;
- Summary of the results obtained;
- Conclusions

6.2. Interview script

The script for the interview had the following questions:

- What are the reasons why the University decided to join EQUIS?
- Who were the actors involved in the EQUIS process? (directors, teachers, students, staff, alumni, business recruiters, other clients).
- Did the actors participate readily? Where was there most resistance?
- What is the role of each of these actors in the implementation of EQUIS?
- With EQUIS, was there improvement in terms of teams?
- With EQUIS, were there changes to the University management?
- What are the areas in which the Faculty has most difficulty in implementing the EQUIS criteria?
- In terms of EQUIS, can one say that there is a set of best practices which are transmitted by people?
- With EQUIS, have there been changes in the sharing of knowledge in the team?
- Is there an Intranet where the processes are managed?
- What kind of innovations have emerged with EQUIS?
- Does the entire University staff have training?
- How is the know-how of employees transmitted? Orally? Written? Have there been changes with EQUIS?

- Does each process (e.g. the creation of a new course) follow a methodology which is described within EQUIS?
- Is the University well aware of its customers? How evident is it? How does the University try to meet the needs of its customers?
- Does the University deal with complaints? When there is some kind of complaint, how is it treated? Is it written down?
- Are there partnerships with various organizations? Are they the same as those that existed before EQUIS or improvements were made? What kind of partnerships are these?
- Does the University share knowledge in the network of EQUIS organizations? What are the advantages?
- Do you consider that EQUIS has made processes simpler or more bureaucratic?
- Do students recognize the added value of EQUIS Accreditation?
- Is the leadership of EQUIS Accreditation process important? Should this leadership be collective or individual?

6.3. Summary of conclusions

In this paper we only summarized the findings from empirical research. Detailed results of this research were presented in a previous paper (see Matos, 2008).

Using the Intellectual Capital Model (Matos and Lopes, 2009) as a starting point, we analyzed the content of the interviews carried out in the two universities. The findings are presented below.

From this analysis, we can conclude that EQUIS accreditation, as a brand, has become the base for the creation of value for the education institutions analysed. It strengthened innovation capacity. This cannot be disassociated from the improvement of processes and giving greater value to the relationship with the client.

It was clear that, in each of the cases, EQUIS accreditation did not require significant additional financial efforts besides those inherent to adherence to the accreditation system itself.

We conclude that by following this route to accreditation, there was innovation. We also know the intellectual capital of the education institutions analyzed was energized. This shows that managing and energizing intellectual capital allows for the stimulation of sustainable innovation. In reality, the analysis of EQUIS accreditation, based on bibliographical research and on the interviews that have taken place, has shown evidence that to reach an accreditation process it is necessary to have very well systematized processes, which will be enhanced through a permanent search for excellence. One other piece of evidence is the partnerships, together with a culture of the network, where new information and communication technologies have been introduced.

Another notion reflected in the accredited institutions is the notion of "good practices" also applied to the processes of continuous improvement. It is also verifiable that there is a prestige and attractiveness effect, in which the benefits of the "brand" give the Institution indirect publicity, exempting it from an equivalent financial effort.

We can further state that, even if one does not bring into practice any other modification, the analysis of the practices that have taken place in similar contexts produces previously unexploited potential. This innovation is a result of this analysis, and one can label it as "innovation arising from the process", supported by a "network culture" and by new technologies.

On the other hand, it was the implementation of the accreditation process that led to the increase in performance, which concurs also with our theory about the effect of these processes on the capacity of organizational innovation, meaning that the creation of the accreditation process empowers intellectual capital converting it into an innovation mechanism.

The findings of this research indicate the need to promote and enhance the intellectual capital of organizations and particularly the institutions of higher education, which is one of the ways to generate innovation and competitiveness. The accreditation processes seem to favour this task because they compel a better fixation on the management of intellectual capital.

7. Final conclusion

In various surveys conducted in SMEs (see Matos and Lopes 2008, Matos e Lopes 2009) it is concluded that companies need systems to standardize routines, instill discipline and reduce dispersion.

Thus, the accreditation of intellectual capital management, developed through a simple non-bureaucratic process can be very important, allowing us to create an environment of teamwork and networking, encouraging the sharing of knowledge which is essential to creating a dynamic innovative.

Like the accreditation of higher education, which currently is a benchmark of quality for the courses and universities, accreditation can be an important competitive advantage for SMEs because it guarantees to their partners that they have the capacity to generate relevant, shared knowledge and induce incremental innovation.

Just as the EQUIS system was designed to help prospective students and recruiting companies from one country to identify those institutions in other countries that deliver high quality education for international management, the ICMA system can also be the best guarantee of SMEs, which can be used in their promotion to key partners.

In research we conducted in Portuguese SMEs, we have found that the innovative capacity of some SMEs can be recognized. The use of an accreditation system could be seen as a guarantee of innovation capacity and therefore an important reference for their partners, so accreditation can be considered as a tool for organizational development.

References

AACSB (Association to Advance Collegiate Schools of Business), [Online], Available: http://www.aacsb.edu/accreditation/ [05 December 2009].

AMBA (Association for Masters of Business Administration) [Online], Available: http://www.mbaworld.com/ [05 December 2009].

Andriessen D. (2005) "Implementing the KPMG Value Explorer: Critical success factors for applying IC measurement tools" Journal of Intellectual Capital, Vol. 6, Issue 4, pp. 474 – 488.

Brooking, A. (1996) Intellectual Capital: core asset for the third millennium enterprise. Thompson, Boston.

Edvinson, L. and Malone, M.S. (1997) "Intellectual Capital", New York ,Harper Collins Publishers Inc.

EQUIS – The European Quality Improvement System [Online], Available: http://www.efmd.org/html/home.asp [05 December 2009].

Law No. 1 / 2003-01-6 [Online], Available: http://snesup.terradasideias.net/htmls/_dlds/lei_1_2003.pdf [07 December 2009].

Law No. 125/2004 of 2004-05-31 [Online], Available:
http://www.ipq.pt/customPage.aspx?modid=1076&pagID=1290 [07 December 2009]..

Lopes, A., Matos, F. (2005) "Técnicas de Gestão do Conhecimento – Métodos de Aplicação e Desenvolvimento Empresarial", Associação Empresarial da Região de Viseu - AIRV, Viseu.

Lopes, A., Matos, F. (2006) – "Avaliação do Programa REDE", Gest-in, ISCTE, Lisboa.

Matos F.; Lopes A. (2008) "Intellectual Capital Management - Certification Model" Paper read at 9th European Conference on Knowledge Management, Southampton Solent University, Southampton, 4-5 September.

Matos, F. (2008) "Searching for an Accreditation Model of Intellectual Capital Management", Paper read at 5th International Conference on Intellectual Capital, Knowledge Management & Organisational Learning, New York Institute of Technology, New York, 9-10 October.

Matos F.; Lopes A. (2009) "Intellectual Capital Management – SMEs Accreditation Methodology" Paper read at European Conference on Intellectual Capital 09, INHolland University of Applied Sciences, Haarlem, The Netherlands, 28-29 April.

Roos, Göran and Johan Roos. (1997) "Measuring your Company's Intellectual Performance".Long Range Planning 30 p. 3.

Sveiby, K. E. (1997) "The New Organizational Wealth. Managing & Measuring Knowledge-Based Assets". Berrett-Koehler Publishers, San Francisco.

Exploring the Relationships Between Creativity, Innovativeness and Innovation Adoption

Eric Shiu

The University of Birmingham, UK

First published in The Proceedings of ECIE 2009

Editorial Commentary

Creativity is the use of skill and imagination to produce something new and generate new ideas or, as claimed by Daniel Pink (2005), the result of the use of right-directed thinking (representing creativity and emotion) over left-directed thinking (representing logical, analytical thought). Innovativeness, according to Rogers (1983), is the degree to which an individual or other unit of adoption is relatively eager for adopting new ideas. Links between these concepts associated with innovation adoption are examined from a consumer perspective by Shiu in this chapter.

He studies determinants of innovation adoption and relations between domain-specific innovativeness and global innovativeness. The different aspects of the concepts are addressed as well as various contextual dependencies. This constitutes an influential and well-documented chapter that contributes to a better understanding of the critical components in an innovation adoption model. This approach is particularly relevant considering increasing customer involvement in the innovation process, facilitated by the development of Web 2.0 technologies.

References

Pink Daniel, 2005. A Whole New Mind: Why Right-Brainers Will Rule the Future. Riverhead Trade Publ.

Rogers, E. M. 1983. Diffusion of innovations (3rd ed.). New York: Free Press

Abstract: This exploratory study attempts to explore the relationships between innovation adoption of blu-ray disc players and its potentially direct and indirect determinants including perceptions of innovation attributes, global innovativeness, domain specific innovativeness and creativity. Results show that this innovation adoption is related to relative advantage and complexity, and more related to domain-specific innovativeness than to global innovativeness, while domain-specific innovativeness, rather than global innovativeness, is significantly related to creativity.

Keywords: Creativity, innovativeness, innovation attributes, innovation adoption, innovation

1. Introduction

Much research has been conducted on identifying factors of innovation adoption (Dupagne and Agostino 1991; Kang 2002; Vishwanath and Goldhaber 2003). One set of factors that has been suggested and popularly referred to is people's perceptions of innovation attributes about an innovation of question (Rogers 1995). However, prior research on the prospective influence of these attributes has not been consistent. For example, in a meta-analysis conducted by Tornatzky and Klein (1982), they found compatibility and relative advantage were not always consistently related to the adoption rate in a positive direction, while complexity was negatively related to the rate of adoption. Another study by Leung and Wei (1999) compared the impact of different innovation attributes on the adoption of mobile phones in Hong Kong, and found that such an impact was significant in only the compatibility and observability variables. Wei (2001) extended Leung and Wei's (1999) research longitudinally and noted that only observability continued to produce a significant impact on people's likelihood to adopt. Against such inconsistent findings, we could conclude that Rogers's (1995) five innovation attributes overall are potentially important predictors of innovation adoption, but which ones are significant, and the relative influences of these attributes, could vary depending in part on what kinds of products and markets are being looked into.

Innovativeness has also been suggested as an important set of factors that can mould one's inclination to adopt an innovative product. For example, Goldsmith, Freiden and Eastman (1995) looked into the fashion and electronic products markets, and suggested the likelihood of adoption of new products in these markets is related to fashion and electronic innovativeness respectively, which in turn are related to global innovativeness. Manning, Bearden and Madden (1995), on the other hand, concentrated on the hypothetical effects of two conceptualisations of innovativeness on innovation adoption, and found that consumer novelty seeking is positively associated with early stages of the adoption process, while consumer independent judgement making is related to later stages of the process.

Creativity itself has been claimed as one last frontier in consumer research that has been significantly under researched (Burroughs and Mick 2004). Its linkage to innovativeness has been hypothesised and rationalised (Hirschman 1980), but so far no empirical research has been carried out to identify any such linkage. In view of its potential indirect relationship to innovation adoption via innovativeness, incorporating the creativity variable in innovation research could contribute to the building up of a more comprehensive innovation adoption model.

In view of the research backgrounds outlined above, this study sets up the two research aims as follows:

1.   Assess the relative strengths of the relationship between perceptions of innovation attributes and innovation adoption, between global innovativeness and innovation adoption, and between domain-specific innovativeness and innovation adoption.
2.   Test the possible relationship between creativity and each of the two levels of innovativeness, i.e. global innovativeness and domain-specific innovativeness.

2. Theoretical foundations

2.1. Innovation attributes

A sizable proportion of previous research on consumer innovation adoption has referred to the renowned five innovation attributes, namely relative advantage, compatibility, complexity, trialability, and observability, proposed by Rogers (1995), who in as early as 1962 first suggested that an individual's degree of beliefs in these attributes can significantly predict most of the variance in innovation adoption. Ostlund (1974) echoed by stating that the more positive an individual's perceptions, as measured by the five aforesaid attributes, about an innovation, the greater the probability of its adoption.

Relative advantage is concerned with 'the degree to which an innovation is perceived as better than the idea it supersedes' (Rogers 1995). The economic aspects, social prestige, convenience and satisfaction are important considerations in measuring the degree of relative advantage. Relating this attribute to the blu-ray disc player, the major relative advantages are that it offers greater storage capacity over conventional DVD players by five to ten times, the best available picture resolution and enhanced interactivity (blu-ray disc website 2009).

Compatibility is about 'the degree to which an innovation is perceived as being consistent with the existing values, past experiences, and needs of potential adopters' (Rogers 1995). In order to assess new ideas, people often make use of old ideas as their main mental tools, which can help to reduce the uncertainty attached to the potential adoption of new products. According to the Blu-ray Disc Association website, DVDs of previous non-blu-ray versions are compatible with blu-ray disc players. This means that consumers will not have to worry about their existing non-blu-ray DVD collections when they are considering the purchase of a blu-ray disc player. In addition, increasingly electronic product innovations are often marketed and consumed as an interrelated bundle of new ideas. The adoption of one new idea may lead to the adoption of several others. For example, blu-ray disc players are compatible with HDTVs, which are becoming more and more popular among households in recent years. There are currently in excess of 700 thousand UK households owning a HDTV. They are all potential customers of blu-ray disc players fuelled by the positive compatibility effect of the HDTVs in their homes.

Complexity is 'the degree to which an innovation is perceived as difficult to understand and use' (Rogers 1995). Understandably it is important that an innovation should not be perceived as too complex for its potential adopters. For example, during the 1980s many earlier adopters of the home computer found it very complex to use and therefore a feeling of frustration sinks in. This perceived complexity led to a negative force in the home computer's rate of adoption during the decade until it was then made more user friendly which saw a remarkable rise in its rate of adoption in the 1990s. In the case of the blu-ray disc player, although it has some new features and offers some extra benefits, it does not require new sophisticated skills and the logic used in its operation is more or less similar to conventional DVDs. Therefore, we assume that most people should not find a blu-ray disc player complex to use.

Trialability is 'the degree to which an innovation may be experimented with on a limited basis' (Rogers 1995). Innovations that can be tried are generally adopted more rapidly than those that precludes potential adopters from trying. The main positive effect of trialability is that it can reduce prospective buyers' sense of uncertainty towards the innovative product. For blu-ray disc players, they can be tried in electronic retail stores as well as on a PlayStation 3 console system. Therefore we conjecture that trialability should not be too much of an issue with the blu-ray disc player.

Observability is 'the degree to which the results of an innovation are visible to others' (Rogers 1995). The easier it is for individuals to visualise how the innovation is being used and its benefits, the more likely these individuals are to adopt it. The degree of observability of the blu-ray disc player is likely to be different between people, which could in turn affect their likelihood to adopt the innovation.

2.2. Consumer innovativeness

Past research has conceptualised consumer innovativeness in two primary ways (Im, Bayus and Mason 2003). First is often named as global innovativeness (Goldsmith and Hofacker 1991; Goldsmith, Freiden and Eastman 1995), but it is also referred to as innate innovativeness (Hirschman 1980). It is essentially an innate phenomenon and is widely referred to in psychology to identify innovative characteristics, if any, of individuals (Kirton 1976). It is not readily observable, and represents a highly abstract and generalised personality trait (Im, Bayus and Mason 2003). This was echoed by Goldsmith and Hofacker (1991) who regarded global innovativeness as a personality trait at the highest level of abstraction, which means that it is 'independent of the context or domain in which consumers are located' (Midgley and Dowling 1978). Roehrich (2004), by referring to Steenkamp et al. (1999), defined it as 'a predisposition to buy new and different products and brands rather than remain with previous choices and consumer patterns'.

Several studies (e.g. Hirunyawipada and Paswan) have identified the multiple aspects of global innovativeness, including openness to information processing, willing to change, inherent novelty seeking, optimum stimulation level, and variety seeking, all of which collectively lead to the tendency to acquire novel information and / or adopt new products. Hirschman (1980) and Manning, Bearden and Madden (1995) equated this innovative trait with consumer novelty seeking, which is concerned with 'an inherent desire to seek out novelty and creativity' (Im, Bayus and Mason 2003).

Second is domain specific innovativeness, which is defined as the explication of 'the narrow facets of human behaviour within a person's specific interest domain' (Midgley and Dowling 1993). Unlike global or innate innovativeness whose influences can be evident across a variety of product classes (Im, Bayus and Mason 2003), domain specific innovativeness is about the tendency to acquire new products or related information within a specific domain. This is perhaps the consequence of an interaction between global innovativeness and a strong interest in the product category concerned (Midgley and Dowling 1978; Roehrich et al. 2002).

In sum, in the innovativeness hierarchy, global innovativeness is conceptualised at the broadest level, whereas domain specific innovativeness is more narrowly defined. Because the latter is more narrowly defined, it is conjectured to have better predictive power of an individual's behaviour – in both the actual adoption of innovative products and the propensity to acquire information associated with new products - within the domain concerned (Hirunyawipada and Paswan). This view has been echoed by Goldsmith, Freiden and Eastman (1995) who found that global innovativeness is weakly related to adoption behaviours, while domain specific innovativeness is strongly associated with adoption behaviours of fashion and electronic innovations. However, Foxall (1995) presented another finding supportive of the influence of global innovativeness in specific adoption situations, in that it is positively related to adoption behaviours of high involvement products such as high value electronic products like blu-ray disc players, but not to those of low involvement products such as food and grocery products.

2.3. Creativity

In contrast with Rogers' five innovation attributes and innovativeness, creativity has received much less attention in consumer behavioural studies (Burroughs and Mick 2004). It can be defined as the generation of novel mental content by the individual, generally within the context of problem solving (Guildford 1965; MadcKinnon 1965). As Hirschman (1983) wrote, in order to solve a problem, a person may 'either restructure internal knowledge in a new way or acquire new information from the environment and combine it with internal information'. Both procedures, according to Guildford (1965), involve the generation of novel mental content arising from a novel cognitive combination of information, which is the essence of creativity. As a more creative person is better in generating new ways and new things and making better use of new information, he/she may be more inclined to try new things and get hold of new things. Hirschman (1980) suggested a number of reasons for the potential linkage between one's creativity and innovativeness, by stating that a more creative person could 'be better able to comprehend the functioning of the innovation', 'able to generate several realistic comparisons between innovation and several alternative products', and 'have a well-developed repertoire of problems the innovation could be used to solve'. These suggested reasons are all sensible but so far the plausible relationship between creativity and innovativeness has not been empirically tested.

In this research, we focus on daily life creativity instead of special talent creativity. The former is vested in virtually every human being to a greater or lesser extent, while the latter is gifted to only a tiny proportion of the population. In order to ascertain the reliability of our creativity test results, we conduct two different creativity tests on each respondent, which will be explained in the following section on measurement scales.

3. Measurement scales

A number of measurement scales have to be decided for this research. The first set of scales is Rogers' five innovation attributes. Preliminary ideas were drawn from reading ideas and practices of previous studies that necessitated establishing the same scales, including Moore and Benbasat (1991), Steckler et al. (1992) and Brink et al. (1995). These ideas and practices were then refined through comparing the contributions of different items used for a scale as well as through a pilot study. Subsequently, also taking the questionnaire length into consideration, two items were chosen and adapted for each innovation attribute, which are described in Table 1.

Table 1: Question items for the three sets of scales

I. Rogers' five innovation attributes
1. Relative advantage
Blu-ray offers far better image and sound quality over DVDs.
Blu-ray offers greater storage capacity compared to other similar formats.
2. Compatibility
I have the required equipments to experience the benefits of blu-ray.
Most of my favourite movies are available on the blu-ray format.
3. Complexity
Learning to operate the blu-ray player is easy for me (reverse).
My interaction with a blu-ray player is clear and understandable (reverse).
4. Trialability
Before deciding whether to buy a blu-ray player, I was able to properly try it out.
Before deciding on whether or not to buy a blu-ray player, I would need to use it on a trial basis.
5. Observability
It was easy for me to observe the difference in the quality of movies on blu-ray compared to DVD.
Before deciding on whether or not to buy a blu-ray player, I was able to watch a demo.
II. Consumer innovativeness
1. Global innovativeness
I am reluctant about adopting new ways of doing things until I see them working for people around me (reverse).
I rarely trust new ideas until I can see whether the vast majority of people around me accept them (reverse).
I am aware that I am usually one of the last people in my group to accept something new (reverse).
I must see other people using new innovations before I will consider them (reverse).
I am generally cautious about accepting new ideas (reverse).
I tend to feel that the old ways of living and doing things is the best way (reverse).
2. Domain specific (electronic) innovativeness
In general, I am the last in my circle of friends to know about the latest new electronic entertainment equipment (reverse).
Compared to my friends, I own very little electronic entertainment equipment (reverse).
In general, I am among the last in my circle of friends to buy new electronic en-

tertainment equipment when it appears (reverse).
I know the names of new electronic entertainment equipment before others do.
If I heard that new electronic entertainment equipment was available in the store, I would be interested enough to buy it.
I will buy a new item of electronic entertainment equipment even if I have had little experience with it.

III. Creativity
1. Hirschman's (1983) Three Task Approach
First task - Name as many different examples as you can of the following: Things that fly: Things that are blue: Things that are old: Things that are strong: Things that can talk:
Second task - Name as many different uses as you can for the following: A shoe: A brick: A coin: A pencil: A light bulb:
Third task - Name as many similarities as you can between the following word pairs: A train and a tractor: Milk and meat: Cat and bird: Education and transportation: Government and religion:
2. Burroughs and Mick's (2004) Problem Scenario Approach
Problem scenario: Just suppose that you are going out to dinner one evening. You have just moved into the area to take a new job. It is the annual company banquet held by your new employer and you are probably going to be called up front and introduced to the rest of the company by your new boss. You put on a black outfit and think you are all ready for the dinner when as you go to put on your shoes, you discover they are all scuffed up and the scuffs are definitely noticeable. You go to the utility closet only to discover that you are almost completely out of shoe polish. This is the only pair of shoes you have to go with this outfit and there is really no other outfit you can wear. You have 2 minutes before you must head to the dinner if you are to be on time. Since you live in a residential area, all of the stores in your part of town have already been closed for the evening. You know of one shopping mall that is open but it means an extra 5 miles of driving. Your solution:
IV. Purchase intention

145

Would you like to try this product (blu-ray disc player)?
Would you buy this product (blu-ray disc player) if you happened to see it in a store?
Would you actively seek out this product (blu-ray disc player) in a store in order to purchase it?

The second set of scales is the two levels of innovativeness, i.e. global innovativeness and domain specific innovativeness. Global innovativeness was to be assessed by a six-item scale first developed by Hurt, Joseph and Cook (1977) and popularly used in later studies such as Goldsmith, Freiden and Eastman (1995). Since the test product chosen is the blu-ray disc player which belongs to the electronic product domain, Goldsmith and Hofacker (1991)'s electronic innovativeness scale was chosen as domain specific innovativeness for this research. The scale, also comprising six items, has been validated for its validity and reliability (Goldsmith and Flynn 1992; Flynn and Goldsmith 1993), and has been used since then (Goldsmith, Freiden and Eastman 1995). Please refer to Table 1 for the items representing these two innovativeness scales.

The third set of scales is daily life creativity, which will be assessed by two different established tests. One is Hirschman's (1983) borrowed Three Task Approach, which addresses three types of mental content that a person may be required to generate while solving daily life problems. These are (1) examples of products possessing a given attribute, (2) different uses of products, and (3) the similarities between two or more products. Although Hirschman (1983) posited that this approach is to measure consumer creativity, it seems more appropriate to be referred to as measuring daily life creativity because all the three tasks may fall upon an individual at any time on a day, such as at study, at work, in the kitchen or in the living room, irrespective of whether this individual is a consumer. In addition, this approach was actually used much earlier before by Getzels and Jackson (1962) and Wallach and Kogan (1965) for studying creativity in school children, implying that the original purpose of this approach is to test creativity in a general context rather than in a consumption context. For averaging out purposes, respondents are asked to complete each task five times, i.e. five different product examples, five different product uses, or five different product similarities.

The second test is Burroughs and Mick's (2004) Problem Scenario Approach, which presents a problem scenario about wearing shoes, which we suggest is more about daily life creativity than consumer creativity that Burroughs and Mick (2004) said, to respondents and then elicits their solutions. This approach is all about evaluating the degree of novelty and functionality, the two core components of creativity, possessed by each respondent. Novelty may involve a new use of a specific product, combining two or more products for a specific consumption purpose, or changing the form of a product in order to enhance its performance. Functionality, on the other hand, is concerned with the extent to which a consumption response effectively addresses the problem at hand or improves on an existing solution.

According to Burroughs and Mick (2004), a number of antecedents are expected to determine the degree of consumer creativity of an individual. Two such antecedents are situational involvement and time constraint. Even though initially an individual may not be particularly interested in an activity, when the situation to which the activity is related exerts a considerable implication for that individual, involvement in the activity will be higher and propensity to make use of consumer creativity will then be higher.

The influence of time constraint on consumer creativity is less straightforward. Extreme time constraints can make an individual overly nervous and therefore can stifle consumer creativity. However, restricted access to products and markets arising from reasonable time pressures, by inhibiting conventional responses, can enhance consumer creativity (Ridgeway and Price 1991).

Accordingly, this research reminded respondents that they are facing a highly involved situation as the dinner is a banquet with the new employer, and also that they have only 2 minutes to tackle the given problem scenario. The two-minute constraint was used because it is short and demanding but could be just enough for respondents to extract their creative thoughts.

All the three sets of scales were pilot tested with five respondents, who were asked to indicate if any of the question items was confusing or difficult to respond. As a result of the pilot test, some question items were reworded, which were then tested again on the respondents who raised the concern. This rewording process was stopped until all the five respondents did not find any problem with the question items.

The fourth set of scales is to elicit respondents' intention to purchase the product of question, i.e. blu-ray disc player. We use a simple three-item scale. This purchase intention scale does not contain too many items and is easy to understand, and it still enables us to test its reliability.

4. Sampling and data collection

The sample would comprise solely university students for two methodo-logical reasons. First is that young students are one of the largest potential market segment of electronic products. In general they tend to be inter-ested, experienced and knowledgeable in electronic products including the blu-ray disc player and therefore their responses to this research could be more thoughtful, accurate and reliable than an average person in a society. Also this research is exploratory in nature and aims at arriving at a hypo-thetical expanded model that includes new extraneous variables. For this kind of exploratory model building research, external validity is less of a concern, and a homogeneous sample such as a student-only sample is deemed better than a heterogeneous sample for minimising the unwanted effect caused by the difference among the sample respondents (Calder 1981). One such difference is the difference in ages between respondents, which has to be controlled in this research because there is a well-accepted perception that older people are less likely to purchase and use new technological products. Numerous previous quality research in the marketing field, such as Boulding et al. (1992) and Boulding et al. (1993) also used students to obtain more homogeneous samples for theory test-ing purposes.

Fifty student-respondents were randomly drawn from a typical civic uni-versity in England. Obviously a sample size of fifty is small. However, be-cause the creativity questions, which demanded much more interviewer and respondent time than typical questionnaire questions, required the interviewer to meet each respondent in person in order to observe the time constraint, they were posed to respondents in a way that is more similar to an in-depth interview situation than a standardised question-naire situation. Such a practice of small sample field work that required both face-to-face creativity tests on and questionnaire survey with each respondent is not uncommon in psychological research. In addition, any statistically significant result derived from a small sample will be more robust than a similar result obtained from a large sample because the effect size factor is less problematic with a smaller sample size.

5. Validity and reliability

All the four sets of scales described above and used in this research were derived from previous studies which had established their internal validity. They were all developed within the context of the US. This research was done in the UK, which is culturally and ethically more similar to the US than to many other countries. We don't expect the transferability of these scales to the UK context to be significantly problematic.

Concerning reliability, we have conducted Cronbach's alpha or Pearson correlation tests on each set of scales. Usually a Cronbach's alpha or Pearson correlation coefficient of 0.7 or above is often the minimum required to ascertain the reliability of the scale concerned. However, considering the small sample size of 50, we would accept any scale achieving a reliability coefficient of around 0.6.

As shown in Table 2, only two of the five Rogers' innovation attributes have achieved a correlation coefficient sufficient for us to suggest that they are reliable. They are 'relative advantage' (0.59 at 0.01 significance level) and 'complexity' (0.73 at 0.01 significance level). The 'unreliability' of the other three scales may be due to the small sample size, small number of items or other unexpected reasons. Whatever the reasons, we will keep only the two reliable scales to represent respondents' perceptions of the innovation attributes of the blu-ray disc player. On the other hand, the innovativeness and creativity scales have passed the reliability tests.

Table 2: Reliability tests of the scales used

Scale	Reliability coefficient	P value
I. Rogers' five innovation attributes		
1. Relative advantage	0.59	0.00
2. Compatibility	0.18	0.21
3. Complexity	0.73	0.00
4. Trialability	-0.06	0.68
5. Observability	0.25	0.08
II. Innovativeness		
1. Global innovativeness	0.67	

Scale	Reliability coefficient	P value
2. Electronic innovativeness	0.82	
III. Creativity		
1. Hirschman's (1983) Three Task Approach		
First marker's assessment	0.85	
Second marker's assessment	0.83	
Correlation between first marker's and second marker's	0.86	0.00
2. Burrough and Mick's (2004) Problem Scenario Approach		
Correlation between first marker's and second marker's	0.35	0.01
IV. Purchase intention	0.702	

Reliability coefficients with a p-value are Pearson's correlation coefficients, while those without a p-value are Cronbach's alpha coefficient

6. Hypothesis development

Rogers posited that consumers' intentions to purchase a new product are affected by the five innovation attributes, namely relative advantage, compatibility, complexity, trialability, and observability. This study has used a two-item scale for each of these attributes, and found two of them, relative advantage and complexity, to be reliable. These two attributes would be used to represent the innovation attributes of the new product, which is the blu-ray disc player in this study. The corresponding hypotheses are:

H1: Purchase intention is positively related to relative advantage

H2: Purchase intention is negatively related to complexity

Goldsmith, Freiden and Eastman (1995) stated that linkages should exist between different levels of innovativeness and that 'global, abstract constructs may be more useful in predicting lower-level abstract constructs than in predicting overt behaviour'. On the premise of this statement, they put forward three levels of innovativeness, i.e. global innovativeness, domain-specific innovativeness, and concrete innovativeness. They then argued that global innovativeness is correlated more highly with domain-specific innovativeness than it is with concrete innovativeness, because the global is closer to the domain-specific than to the concrete in level of abstraction. Their definitions of global innovativeness and domain-specific innovativeness, as well as the items used to represent each of these two scales, are the same as this study. They defined concrete innovativeness as concrete innovative purchasing behaviour in a product field concerned, which is akin to but one step further than the purchase intention scale developed in this study. Following Goldsmith, Freiden and Eastman's (1995) arguments and logics, this study put forward the following five hypotheses:

H3: Global innovativeness is positively related to electronic innovativeness

H4: Global innovativeness is positively related to purchase intention

H5: Electronic innovativeness is positively related to purchase intention

H6: Global innovativeness is more related to electronic innovativeness than to purchase intention

H7: Purchase intention is more related to electronic innovativeness than to global innovativeness

The possible linkage between an individual's level of creativity and his/her level of innovativeness has not been explored in previous research. However, Shiu (2009) posited that innovation is the de facto application of creative ideas. Also as described in the earlier part of this paper, a more creative individual is more able to develop new ways and new things as well as make better use of new information, and therefore he/she may have a higher propensity to try new things and get hold of new things. These new things could be things in general, or things in a particular domain. Along this line of thoughts, the following two hypotheses have been suggested:

H8: Creativity is positively related to global innovativeness

H9: Creativity is positively related to domain-specific innovativeness

On the other hand, we don't expect creativity to be related to concrete innovativeness because apparently the two may be too distant from each other to allow for any clearly explainable relationship.

7. Results

A number of Pearson's correlation tests indicated that most of the hypotheses established above are proved correct. Specifically, hypothesis 1 'relative advantage is positively related to purchase intention' and hypothesis 2 'complexity is negative related to purchase intention' are confirmed because the correlation coefficient is 0.48 for the former, and 0.37 for the latter, both at 0.01 significance level. Hypothesis 3 'global innovativeness is positively related to electronic innovativeness' is correct as the correlation coefficient is 0.47 at 0.01 significance level. Hypothesis 4 'global innovativeness is positively related to purchase intention' is confirmed because the two are significantly correlated at 0.05 significance level and the correlation coefficient is 0.35. Hypothesis 5 'electronic innovativeness is positively related to purchase intention' is also right as the two constructs generate a correlation coefficient of 0.52 at 0.01 significance level.

Comparing the strengths of relationship among the three levels of innovativeness, hypothesis 6 'global innovativeness is more related to electronic innovativeness than to purchase intention' is validated because the global innovativeness – electronic innovativeness correlation coefficient of 0.47 is larger than the global innovativeness – purchase intention correlation coefficient of 0.35. Hypothesis 7 'purchase intention is more related to electronic innovativeness than to global innovativeness' is also right as the purchase intention – electronic innovativeness correlation coefficient of 0.52 is greater than the purchase intention – global innovativeness correlation coefficient of 0.35. Concerning the possible relationship between creativity and innovativeness, the situation is less straightforward than the above confirmed relationships. Results of this study indicate that creativity is positively related to electronic innovativeness, but not related to global innovativeness. Therefore, a more creative person is not likely to be more innovative in a general sense as what we conjecture. However, a creative individual has been found to be more electronically innovative. This implies that a more creative person is more likely than a less creative person to be engaged in new electronic products.

8. Conclusion

This study has achieved a number of research targets. First is about the comparison of the strengths of relationship between innovation adoption and perceptions of innovation attributes represented by relative advantage and complexity, between innovation adoption and global innovativeness, and between innovation adoption and domain-specific (electronic) innovativeness. Results show that domain-specific innovativeness, as expected, is more related than global innovativeness to innovation adoption because theoretically the former is closer than the latter in the hierarchy of innovativeness to innovation adoption. Results also show that this domain-specific innovativeness can account more than perceptions of innovation attributes people's likelihood to adopt an innovative product. Second, we've found that creativity is not significantly related to one's global innovativeness but it is overtly related to electronic innovativeness. In other words, being more creative does not mean that a person is more prone to try new ideas and new things, but he/she is more inclined to try out new electronic products.

The research aims that have been achieved through this study could provide some pioneering insights into future directions in innovation and creativity research. Although the small sample size of 50 used in this study can be regarded as a strength because any significant result obtained from such a small sample size should be more robust than a much larger sample size. However, this can also be treated as a weakness because the small sample size has forced us to include those results whose p values are moderately greater than 0.05 as potentially statistically significant, which sometimes may not be true. In order to eradicate this somewhat arbitrary practice, it is deemed better to have a larger, but not significantly larger, sample size to test all the hypotheses set up in this study. A sample size of 100 or 200 would be a good one. Regarding the use of only the students for our sample, we deem that this is not a major limitation because the use of a more homogeneous sample can reduce the bias arising from the diverse backgrounds if we draw our sample from the total population.

References

Brink, S.G., Basen-Engquist, K.M., O' Hara-Tompkins, N.M., Parcel, G.S., Gottlieb, N.H. and Lovato, C.Y. (1995) Diffusion of an effective tobacco prevention program: I. Evaluation of the dissemination phase, Health Education Research, 10, 283-295.

Burroughs, J.E. and Mick, D.G. (2004) Exploring antecedents and consequences of consumer creativity in a problem-solving context, Journal of Consumer Research, Vol.31, No.2, 402-411.

Dupagne, M. and Agostino, D.A. (1991) High-definition television: a survey of potential adopters in Belgium, Telematics and Informatics, 8(1-2), 9-30.

Flynn, L.R. and Goldsmith, R.E. (1993) Identifying innovators in consumer service
markets, Service Industries Journal, 13(July), p.97-109.

Getzels, J.W. and Jackson, P.W. (1962) Creativity and intelligence, New York: John
Wiley.

Goldsmith, R.E. and Flynn, L.R. (1992) Identifying innovators in consumer product
markets, European Journal of Marketing, 26(12), p.42-55.

Goldsmith, R.E., Freiden, J.B. and Eastman, J.K. (1995) The generality/specificity
issue in consumer innovativeness research, Technovation, 15(10), p.601-612.

Goldsmith, R.E. and Hofacker, C.F. (1991) Measuring consumer innovativeness,
Journal of the Academy of Marketing Science, 19(Summer), p.209-221.

Guildford, J.P. (1965) Intellectual factors in productive thinking, In: Productive
Thinking in Education, Washington, D.C.: National Education Association, p.5-
20.

Hirschman, E.C. (1980) Innovativeness, novelty seeking, and consumer creativity,
Journal of Consumer Research, Vol.7, No.3, 283-295.

Hirschman, E.C. (1983) Consumer intelligence, creativity, and consciousness: implications for consumer protection and education, Journal of Public Policy and
Marketing, Vol.2, Issue 1, 153-170.

Hurt, H.T., Joseph, K. and Cook, C.D. (1977) Scales for the measurement of innovativeness, Human Communication Research, 4(Fall), p.58-65.

Im, S., Bayus, B.L. and Mason, C.H. (2003) An empirical study of innate consumer
innovativeness, personal characteristics and new product adoption behaviour,
Journal of the Academy of Marketing Science, Vol.31, No.1, 61-73.

Kang, M.H. (2002) Digital cable: exploring factors associated with early adoption,
Journal of Media Economics, 15(3), 193-207.

Kirton, M. (1976) Adaptors and innovators: a description and measure, Journal of
Applied Psychology, Vol.61, No.5, 622-629.

Leung, L. and Wei, R. (1999) Who are the mobile phone have-nots? Influences and
consequences, New Media and Society, 1(2), 209-226.

Manning, K.C., Bearden, W.O. and Madden, T.J. (1995) Consumer innovativeness
and the adoption process, Journal of Consumer Psychology, 4(4), 329-345.

Moore, G.C. and Benbasat, I. (1991) Development of an instrument to measure the
perceptions of adopting an information technology innovation, Information
Systems Research, 2(3), 192-222.

Ostlund, L.E. (1974) Perceived innovation attributes as predictor of innovativeness,
Journal of Consumer Research, 1: 23-29.

Pankratz, M., Hallfors, D. and Cho, H. (2002) Measuring perceptions of innovation
adoption: the diffusion of a federal drug prevention policy, Health Education
Research, Vol.17, No.3, 315-326.

Rogers, E.M. (1995) Diffusion of Innovations, 4th edition, New York: Free Press.

Steckler, A., Goodman, R.M., McLeroy, K.R., Davis, S. and Koch, G. (1992) Measuring the diffusion of innovative health promotion programs, American Journal of Health Promotion, 6, 214-224.

Vishwanath, A. and Goldhaber, G.M. (2003) An examination of the factors contributing to adoption decisions among late-diffused technology products, New Media and Society, 5(4), 547-572.

Wallach, M.A. and Kogan, N. (1965) Modes of Thinking in Young Children, New York: Holt, Rinehart and Winston.

Wei, R. (2001) From luxury to utility: a longitudinal analysis of cellphone laggards, Journalism and Mass Communication Quarterly, 78(4), 702-719.

Creativity and Discovery: Moving Beyond Market Equilibrium

Greg Clydesdale

Massey University, Auckland, New Zealand

First published in The Proceedings of ECIE 2009

Editorial Commentary

The pursuit of opportunity is at the core of the entrepreneurship concept. The fundamental concern for an entrepreneur is to identify the right set of circumstances to undertake entrepreneurial activities (Timmons, 1994). Identifying and selecting opportunities for a new business is among the most important abilities for a successful entrepreneur (Stevenson, 1991; Ardichvili & Cardozo, 2000, p.17).

The research authored by Clydesdale is well anchored conceptually and interestingly questions the traditional meaning of, "opportunity discovery and opportunity creation". It proposes new definitions that are located in a creativity/discovery spectrum which recognizes that opportunities can pre-exist, or they can be created. It also introduces the notion of threshold – a point at which an opportunity comes in to existence. This represents a serious and original contribution to the field of entrepreneurship and market dynamics, providing a framework consistent with existing theory which unveils new ways to consider entrepreneurial opportunities. The treatment of the topic similarly reflects the insights developed in the paper.

References

Ardichvili, Alexander; Cardozo, Richard N, 2000. A Model of the Entrepreneurial Opportunity Recognition Process. Journal of Enterprising Culture, Jun2000, Vol. 8 Issue 2, p103, 17p

Timmons, JA, Muzyka, DF, Stevenson, HH, & Bygrave, WD 1987. Opportunity recognition: The core of entrepreneurship. In NC Churchill, JA Hornaday, BA Kirchhoff, OJ Krasner, & KH Vesper (Eds.), Frontiers of entrepreneurship research, pp. 109-123.

Abstract: This paper argues that an emphasis on supply, demand and equilibrium has constrained entrepreneurial education. In reality, entrepreneurs may be involved in substantial non-market activity to bring an opportunity into existence. In contrast to definition of opportunity and discovery which emphasizes supply and demand, a simple definition of opportunity creation and discovery is proposed that accommodates important non-market activity. Opportunity discovery is defined as finding a pre-existing opportunity to create a sustainable business. By contrast, with opportunity creation, the opportunity does not exist and environmental modification is necessarily to bring the opportunity into existence, after which a sustainable business can be built. An example of opportunity creation through non-market activity is provided, in particular management of government relations. Consistent with these definitions, two important concepts are developed. First is the Creativity/Discovery spectrum which recognises that each opportunity has its own particular combination of creation and discovery activities depending on the work an entrepreneur must do to bring the opportunity into an existence. The second is the concept of thresholds, the point at which an opportunity comes in to existence. An entrepreneur may have to develop necessary environmental forces past a threshold after which the environment will support the business. If the threshold is already passed, the opportunity is ready for discovery. Thresholds can partly explain phenomena such as Kirzner's errors of optimism and pessimism, and contributes to our understanding of timing, a much over-looked factor of entrepreneurship. Moving further away from notions of supply and demand, the Product Life Cycle is developed as an alternative to the equilibrium model in explaining the appearance of opportunities. It draws on previous research to illustrate the role of niche creation and closure for business opportunity.

Keywords: Opportunity, creativity/discovery spectrum, threshold, product life cycle, equilibrium

1. Introduction

Market equilibrium remains a corner-stone of entrepreneurship. It has been linked to source of opportunities (Drucker 1985). Opportunities have been further categorised as Kirznian opportunities which take the market to equilibrium and linked to discovery, or Schumpeterian opportunities which take the market away from equilibrium and linked to creativity.

An emphasis on reading market forces is a logical place to teach opportunity identification, and is intrinsically linked to the issue of whether opportunities are created or discovered. Baker and Reed's (2005) research suggests that most entrepreneurial opportunities were created by entrepreneurs, not pre-existing conditions in the market. In sharp contrast, population ecology theory stresses that the birth rate of firm is determined by environment forces (Hannan and Freeman 1984). Structuration theory is one way of accommodating this impasse, as it sees the entrepreneur and environment evolving together (Sarason, Dean, & Dillard 2005).

The literature on opportunities is still limited, with little for example, on the importance of timing, yet in layman's terms, success is often seen as being in the 'right place at the right time'. This article attempts to contribute to this literature on opportunity creativity and discovery. This paper attempts to resolve the creativity/discovery debate by saying opportunities can be placed on a spectrum depending on the extent to which the entrepreneur brought the opportunity in to existence. It does not define all opportunities as either creation or discovery-based, but argues that each opportunity embodies differing degrees of creation depending on how much work is needed to bring the opportunity into existence.

This paper also argues that the debate needs to be freed from its emphasis on supply, demand, equilibrium and market activity. Insufficient recognition is given to non-market forces important for the creation and exploitation of opportunities. Opportunities can be derived through substantial non-market activity.

The next section discusses the interpretation of opportunity creation and discovery in terms of supply/demand and equilibrium, and argues that this market focus down-plays the importance of non-market activity. It provides simple definitions of opportunity creation and discovery in contrast to those that emphasise market activity. Section three then provides an example of opportunity creation through non-market activity, in particular, management of the political process. A key concept for the existence of opportunities is thresholds, when an environmental force reaches a point that sustains the existence of a business. When a threshold is passed an idea becomes an opportunity. Thresholds also provide some understanding of timing for business venture success. The last section then suggests that the Product Life Cycle can be extended to provide a better alternative to the equilibrium model in explaining the appearance of opportunities.

2. Opportunities and equilibrium

Although Schumpeter is constantly aligned with the idea of opportunity creation, he does not mention it in his text at all. He speaks instead of the creation of new combinations of means of production, which he equates to the production-function (Schumpeter 1934). These combinations may be the case of the introduction of a new good or new method of production, the opening of a new market, the conquest of a new supply of raw materials or the carrying out of a new organisation form. The entrepreneur creates a new combination and pursues it in the market and this has been interpreted as the creation of an entrepreneurial opportunity.

Schumpeter believed that opportunities rarely exist as a result of changes in consumer behaviour. This provides some room for juxtaposing his position with those who argue that opportunities exist in the market place waiting to be discovered. While Schumpeter does not speak of opportunities, he does speak of 'possibilities'. He states "new possibilities are constantly being offered by the surrounding world, in particular new discoveries are continually being added to the existing store of knowledge..." (Schumpeter 1934, p.79). For Schumpeter, it is not a question of discovering but a question of doing. When no one is doing, "possibilities are dead". From an economic perspective, inventions are irrelevant. It is the bringing of the invention into the market that is economically important, and that is the role of the entrepreneur. In this light, the notion of opportunity creation linked to Schumpeter is limited to the economic market sphere. As Buenstorf states:

> ...while the Kirznerian entrepreneur discovers and pursues opportunities that exist within markets (and are reflected in the price system), the Schumpeterian entrepreneur discovers opportunities that exist outside the economic sphere (and are not yet reflected in the

*price system) and pursues these opportunities by bringing them into
the marketplace....*

Subsequent authors have also framed opportunity discovery and creation in terms of market forces. Sarasvathy et al. (2003) present 'opportunity recognition' as a situation in which both sources of supply and demand already exist, and the entrepreneur recognising this, matches them up. They present 'opportunity discovery' as a situation where on one side, supply or demand, exists and the other needs to be discovered. Finally, 'opportunity creation' is where neither supply nor demand exists and both need to be created.

Supply, demand and the notion of equilibrium have been at the heart of entrepreneurship. Dean and Meyer (1996) show a strong correlation between the rate of growth in demand and growth in the number of new businesses. On the other side, supply-driven opportunities are fuelled by changes in technology, resource prices, changes in the number of sellers, and changes to other industries that use the same resources. However, the equilibrium model has been criticised by Shane and Eckhardt (2003) who identify a number of weaknesses with the model including the underlying assumption that prices convey all the relevant information. They also criticise the assumption that all information and expectations about the future can be reduced to current price bids.

Entrepreneurs rely on a number of environmental forces, not just supply and demand, and entrepreneurs may be involved in substantial non-market activity to bring an opportunity into existence. Defining opportunity creativity and discovery in terms of market activity is too limiting and its focus on market activities is too narrow. Schumpeter certainly saw a role of the entrepreneurial leader as having to over-come non-market barriers including legal and political impediments. Opportunities emerge from a complex pattern of changing conditions including technological, economic, political, social, and demographic conditions (Baron 2006). It is the juxtaposition or confluence of conditions at a given point of time that determine the existence of an opportunity.

Buenstorf (2007) argues that the vast majority of entrepreneurial opportunities are created by human activity rather than exogenous forces (eg: natural disasters). This can include significant non-market activity through such things as new inventions and scientific discoveries. For Buenstorf, the issue is not whether the opportunity pre-exists or is actively created, but the extent to which an entrepreneur brings the opportunity in to existence. He distinguishes between opportunities, and 'higher order opportunities' ie "an opportunity to create the conditions for an entrepreneurial act by means of some targeted activity". When one discovers a higher order opportunity, s/he discovers and opportunity to create an entrepreneurial opportunity. An entrepreneur can 'discover' a 'higher order opportunity and 'creates' the conditions for business success.

An opportunity may occur as a result of un-intended human activity. They may also arise from the entrepreneur's own activity even though s/he might not necessarily be pursuing an entrepreneurial idea, for example, an academic researcher may make a discovery for scientific research but it has the added unintended effect of creating a business opportunity.

Drawing on these insights, I propose two simple descriptions of opportunity creation and discovery. Opportunity discovery is defined as finding a pre-existing opportunity to create a sustainable business. By contrast, with opportunity creation, the opportunity does not exist and environmental modification is necessarily to bring the opportunity into existence, after which a sustainable business can be built. By this definition, recognition is not a market state as per Sarasvathy et al (2003), but a state of higher alertness in the individual.

3. Creating opportunities through non-market activity

One of the key environmental forces that can affect the existence of opportunities is government policy. Government policy can provide many opportunities through the large resources at their disposal, and their rules determining how markets operate. When laws are passed affecting what consumers can and cannot buy, they are effectively changing consumer tastes by decree. Changes in how governments manage their own resources can signal opportunity. The different approaches that businesses take towards government policy can help us distinguish the difference between opportunity creation and discovery. It can also provide further insight in to how businesses can manage their environment to create opportunities.

Weidenbaum (1980) outlined three general business responses to public policy. Although not linked to opportunity creation, Weidenbaum's classification can be used as a basis to distinguish between opportunity creation and discovery. Weidenbaum argues that businesses have three approaches they can take with regard to public policy. The first is 'passive reaction' in which a business simply reacts to government policy as it occurs. The second is positive anticipation in which the business is aware that changes are coming and anticipates these changes when they make their business strategies. The third approach is 'public policy shaping' in which the business becomes actively involved in the formation of policy and seeks to shape the political outcomes so that they are aligned with their own interests.

In the first two approaches, the business plays no direct role in the public policy process. If a firm takes either of these stances, it is doing nothing to shape the environment or create the opportunity. If the business takes a 'passive reaction' approach it waits until the changes are introduced and then acts so that they benefit from any opportunity that results. Of course, because they have been so passive, it is possible that the legal outcome may not have a favourable outcome. It may not result in an opportunity, but if it does, the business 'discovers' the opportunity and acts accordingly. In the second option 'positive anticipation', the firm is more positive that a favourable outcome will result and plans for that outcome. It has 'discovered' an opportunity will open in the future, and has created strategies to exploit that opportunity when it finally comes.

In the third option 'public policy shaping', businesses are more pro-active. When a business actively attempts to shape public policy, it can help define laws and regulations in their best interests. This includes removing barriers to markets, or opening markets, perhaps by creating laws about product use. In this way, entrepreneurs can use the public policy process to create opportunities.

There are a number of approaches that an entrepreneur can utilise when attempting to shape government policy. Hillman and Hitt (1999) suggest three generic strategies. The first is 'information strategy' in which an entrepreneur provides government decision makers information through lobbying, research reports and other techniques, in an attempt to shape their decision. The second political strategy is 'Financial incentive strategy' in which the entrepreneur targets political decision makers with financial inducements. The final strategy is 'Constituency-building strategy' in which the entrepreneur works to gain the support of individual voters and citizens, who in turn seek to influence political decision makers.

An excellent example of opportunity creation through political activities can be seen in the actions of Microwave Communications Inc (MCI) who used political activity to open a business opportunity in the long distance telephone market (Yoffie and Bergenstein 1985). This market was regulated by the Federal Communications Commission (FCC) who had given AT&T a monopoly. It was commonly believed that long distance telecommunications transmission was a 'natural monopoly' whose high capital costs meant the best structure was to have only one provider.

MCI wanted to enter the industry using new microwave technologies, however both AT&T and FCC opposed opening the market. It was founded in 1963, by an entrepreneur who intended to compete using microwave technology. To do this, it needed to get approval from the FCC but found itself in a battle with one of America's biggest corporations. Between 1963 and 1969, it was entrenched in legal and political warfare, but generated no revenues.

In 1968, William G. McGowan was appointed CEO, and moved the company's headquarters to Washington, D.C; an indication that political management was central to the company's success. He developed contacts, testified in Congressional hearings, pleaded MCI's case to the FCC, and publicized MCI where ever possible.

Officers in the government receptive to deregulation and competition were targeted and became welcome allies in his battle. His arguments were couched in economic terms that stressed the benefits to customers, and the role of competition in stimulating technological innovation. The debate was also couched in terms of broader social goals, in particular concerns of economic advancement and consumer welfare. In 1976, MCI joined together with a few other small companies to establish the Ad Hoc Coalition for Competitive Telecommunications (ACCT) to help in lobbying. A final strategy involved use of the courts, such as an anti-trust suit it took against AT&T in 1974.

Through aggressive political and legal action McGowan managed his political environment in a manner that successfully opened a business opportunity. The regulatory change that he championed was a 'threshold' that opened the opportunity he sought..

4. Thresholds

In some cases, an entrepreneur introducing a new product knows demand exists before the product is created. For example, demand exists for a car battery that doesn't go flat, but supply is not available due to technological limitations. This shows that we should not over-emphasise the importance of new ideas. Sometimes opportunities stem from old ideas that suddenly become feasible. The difference between an idea and opportunity is its feasibility (Timmons, 1999), and this is often determined by forces beyond our control.

To explore 'feasibility' and opportunity we should consider the concept of 'thresholds'. A threshold is a point that is passed when an environment force reaches a level that is sufficient to support a business. At a threshold, an idea turns into an opportunity. A given state of technology and market size is necessary for a product or service to be sustained. Once a demand or technology threshold is met an idea becomes feasible and an entrepreneurial opportunity is opened. Technological thresholds need to be passed to exploit demand. Demand thresholds need to be passed to exploit technology. Companies with significant marketing and R&D resources have greater potential to manipulate these forces, but the small entrepreneur is more dependent on the initial state of the environment. Thresholds are not just restricted to technology and demand. Thresholds can be found in other environmental forces such as levels of government and infrastructure.

If demand has not reached the necessary threshold, a business will suffer. Thresholds can be paired with Kirzner's work on entrepreneurial errors. Kirzner (1998) described errors of optimism and errors of pessimism. Errors of pessimism provide opportunities. These occur when people in the market place believe something can't be done, when in reality it can. Opportunities exist in the market place and are just waiting to be seized. This would occur when the thresholds have been passed but no entrepreneur has yet noticed their passing. By contrast, errors of optimism occur when entrepreneurs believe something can be done but it can't. In other words, an entrepreneur starts a business before the threshold has been reached. The entrepreneur makes an error of optimism believing the market and other environmental features are sufficient, but they have moved too early.

164

Thresholds are not the only cause of errors. If the threshold passing is noted by a number of entrepreneurs, the resulting competition could reduce the attractiveness of the opening. Another danger signalled by thresholds is that of moving too early.

The concept of thresholds can give a new angle from which to explore the distinction between opportunity discovery and creation. An opportunity might exist once a threshold is passed, but can sit un-observed for some time before it is discovered. It can be discovered by search or accident. By contrast, opportunity creation could be said to occur when an entrepreneur acts to push forces past that threshold in the way that MCI actively lobbied to push the environment past the regulatory thresholds to create an opportunity.

The debate over whether opportunities are discovered or created is one in which opportunities either pre-exist in the environment or come in to being as a result of the entrepreneurs actions. In this light, entrepreneurial opportunities could be placed on a spectrum which at one end has opportunities already existing, waiting to be discovered, while at the other end, are those opportunities that must be brought in to creation by the entrepreneur. We can see this in figure. 1 which illustrates to what extent the entrepreneur brings a particular opportunity into existence.

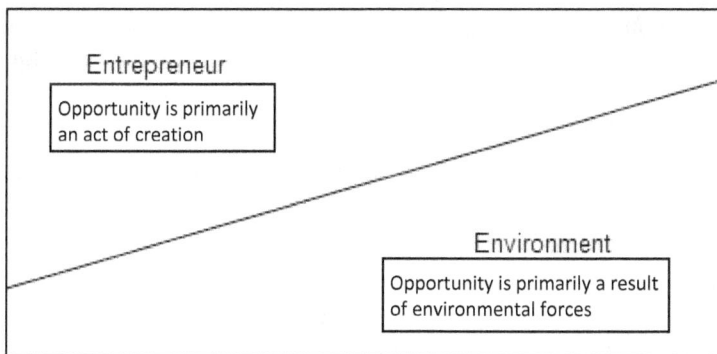

Figure 1: Creativity/discovery spectrum

Those entrepreneurs operating at the right hand of the spectrum have discovered an opportunity which environmental conditions have deemed favourable for the up-take. The technology, market and political-legal situation are all favourable. An entrepreneur in this case will discover, evaluate and exploit the opportunity (Shane and Venkataraman 2000). It is here we find the type of entrepreneurs identified by Shane (2000) who when presented with knowledge of a technological change knew from their previous experience, that the conditions were right for their product. Similarly, the entrepreneur who searches for ideas could also be found here (Hills and Schrader 1998).

On the left hand of the spectrum, environmental forces are not yet sufficient and the entrepreneur needs to work on these before the opportunity is created. This may mean demand is not yet sufficient, and market development is needed to raise demand to a sustainable threshold. Perhaps the technology needs further development, or some lobbying is required to create a legal change. However, even a creative entrepreneur is dependent on the environmental pre-conditions. In between these two extremes, are the numerous variants of entrepreneurship that include both an element of discovery and creativity.

Although creative entrepreneurs have an extra layer of work, this does not necessarily mean that the creative entrepreneur is the hardest worker. Discovery can be preceded by search and a discoverer may have spent three years searching for opportunities before finding one. On the other hand, someone with good political connections may find it easy to create a legal change that brings an opportunity into existence. There is also a lot of work to be done on the exploitation of the opportunity, mobilising the resources and bringing the venture to fruition.

The following propositions are based on the assumption that all opportunities are dependent, in varying degrees, on the state of the business environment:

Proposition.1. More discovery processes occur where the opportunity pre-exists in the environment, while more creativity occurs where the environment needs to be modified to bring the opportunity into existence.

Proposition.2. Thresholds are key points in which environmental forces reach a level where an opportunity comes into existence.

Proposition.3. Successful timing requires an entrepreneur to act after the threshold is reached. Entrepreneurs acting before the threshold is reached will either fail, or if resources are available, complete a process of opportunity creation.

5. Market dynamics and entrepreneurial opportunity

In contrast to the equilibrium model, the Product Life Cycle (PLC) may provide a stronger alternative for illustrating market dynamics and the appearance of opportunity. Timmons (1994, p.91) used the PLC to explain windows of opportunities, while Low and Abrahamson (1997) suggested different types of entrepreneurs are better suited for different stages of the life cycle. This paper extends the model to show changing market dynamics over time. This model can be integrated with the concept of thresholds to explain the importance of timing.

The first stage of the product life cycle is one in which a new product enters the market. In the growth phase, more consumers recognise the benefit of the product and the market begins to take-off, opening opportunities for imitators. This stage is the classic window of opportunity (Timmons 1994) attracting entrepreneurs and venture capitalists with superior returns and the lower rate of failure. More competitors can increase the industry marketing effort and help educate potential customers to the benefits of the product.

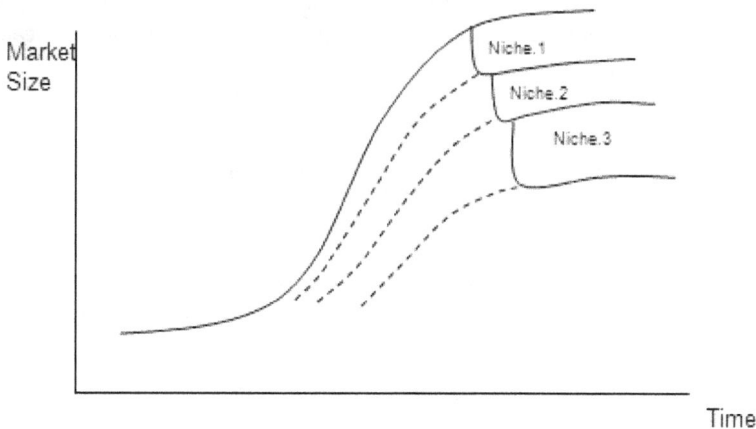

Figure 2: Fragmentation of markets into niches

A number of writers have mentioned the appearance of niches as markets evolve, for example Klepper (1997). Such niches present opportunities for small businesses who cannot compete with large ones who exploit economies of scale. The PLC can be adapted to illustrate niche development. In the early stages of the cycle, small numbers of consumers might prefer a variation of the product. These people are forced to either buy the product as it is, or not buy it at all. As the market grows, the number of people in this category at some point crosses a numerical threshold in which demand justifies making a product tailored to their needs. This threshold heralds the arrival of a sustainable niche. This is illustrated in figure.2, the dotted lines represent that section of the market consuming the general product, which becomes a sustainable niche when the black line forms.

With the crossing of a threshold, an opportunity is born. Entrepreneurs may be alerted to its existence in a number of ways including market research or frequent requests by customers. The importance of timing is illustrated in this model. An alert entrepreneur who identifies the rising market, and sets up business before the market is of sustainable size, will encounter insufficient demand and struggle. By contrast, an entrepreneur who enters the market after that niche is opened has sufficient demand.

The emergence of a new niche signals an opportunity that would appear to favour the small entrepreneur over larger firms. Emerging niches are typically small and favour firms that do not operate with economies of scale. They also favour businesses that can move quickly in response to the changing market (Shane 2000, and Dorfman 1987). However, niches are not only driven by demand, but also technology. If the products characteristics lend themselves to economies of scales, it may be hard to develop a competitive niche. On the other hand, markets can reach a size where they too can exploit scale.

Proposition.4 As some markets grow, niches come into existence of a size that can support a business, presenting an opportunity for discovery.

Proposition 5. If the niche has not reached sufficient size, the entrepreneur has to proportionately increase the work in developing the market and creating the opportunity.

As figure.3 illustrates, niche openings also exist in mature markets as entrepreneurs gain more knowledge of the values consumer seeks, and the fluctuation of market forces. The greater the sales dynamics of industry niches, the greater the formation of new ventures (Dean and Meyer 1996).

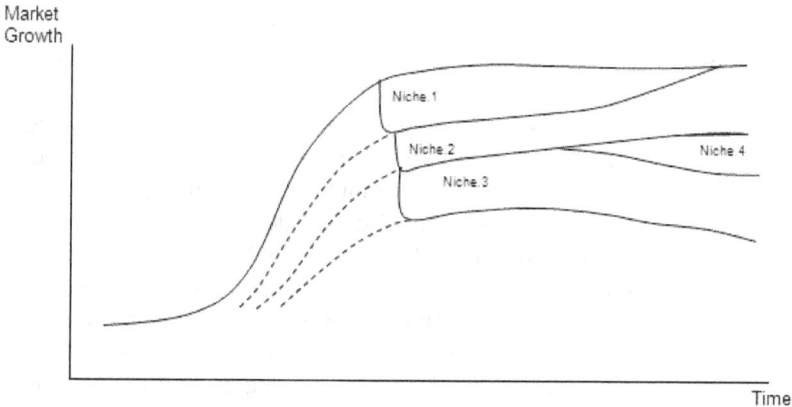

Figure 3: Niches open and close over time

6. Conclusion

This paper argues that entrepreneurial study can be advanced if the binds of market forces are broken. In contrast to definition of opportunity and discovery which emphasises supply and demand, new definitions are provided that are simpler and more amenable to non-market activity. This is followed by an example illustrating non-market activity in the creation of an opportunity. A Creativity/Discovery spectrum is proposed that recognises that opportunities may pre-exist or be created. Opportunities exist on a spectrum depending on the extent of human activity required to bring them in to existence. Central to this is the notion of 'thresholds' a point in which an opportunity comes in to existence. The difference between an idea and an opportunity is its feasibility and that is frequently determined by environmental forces necessary to support a business idea. Thresholds lend themselves to the PLC, Kirznian errors, and a more distinct definition of creativity and discovery. Thresholds increase our insight into the importance of timing and luck and heighten consideration of non-market forces that need to be passed to bring an opportunity in to existence. In recognising the importance of non-market activity, a threshold view of opportunity creation provides a more complete view of the entrepreneurial process. In a further attempt to breach the limitations of supply, demand and the equilibrium model, the Product Life Cycle is developed to illustrate the emergence of niches and the importance of thresholds. In this way, it provides an important educational tool to explore market dynamics that provide entrepreneurial opportunity, and aid the entrepreneur with identification. These dynamics include growth stages, niche development, thresholds, the importance of timing and luck. At the moment the model has several limitations. For example, there are significant variations in the PLC, but these variations do not undermine the model's value. In fact, variations can be explored to enhance student's knowledge of market dynamics and opportunity appearance. The model also provides a model for industrial historians to illustrate changing markets over time. The extent to which the framework is compatible with existing theory, along with its simplicity, makes it an attractive tool. This framework is particularly valuable in teaching opportunity identification. If students can analyse these in different industries they can develop greater understanding of how opportunities are shaped by environmental forces. It suggests a greater role for classwork on longitudinal market evolution. It also suggests a need for research into different types of 'windows of opportunities' and more longitudinal studies on how specific industries and markets change over time, and how opportunities have emerged within them.

References

Baker, T. and Reed, E.N. (2005) "Creating something from nothing: Resource construction through entrepreneurial bricolage", Administrative Science Quarterly, Vol.50, No.3, pp 329-366.

Baron, R. A. (2006) "Opportunity Recognition as Pattern Recognition: How Entrepreneurs Connect the Dots to Identify New Business Opportunities", Academy Of Management Perspectives, Vol. 20, No.1, pp104-119.

Buenstorf, Guido. (2007) "Creation and Pursuit of Entrepreneurial Opportunities: An Evolutionary Economics Perspective", Small Business Economics, Vol.28, No.4, pp323-337.

Dean, T.J. & Meyer, G.D. (1996) "Industry environments and new venture formations in U.S manufacturing: A conceptual and empirical analysis of demand determinants", Journal of Business Venturing, Vol.11, No.2, pp107-132.

Dorfman, N. (1987) Innovation and market structure: Lessons from the computer and semi-conductor industries, Ballinger Publishing Company, Cambridge MA.

Drucker, P. (1985) Innovation and entrepreneurship, Harper and Row, New York.

Hannan, M.T and Freeman, J. (1984) "Structural inertia and organizational change", American Sociological Review, Vol.49, No.2, pp149-164.

Hillman, A. J. and Hitt M.A.(1999) "Corporate Political Strategy Formulation: A model of approach, participation and strategy decision", Academy Of Management Review, Vol.24, No.4, pp825-842.

Hills, G.E. and Shrader, R.C. (1998) "Successful entrepreneur's insights into opportunity recognition", Frontiers of Entrepreneurship Research, Vol.18, pp30-41.

Kirzner, I.M. (1998) How markets work: Disequilibrium, entrepreneurship and discovery. Occasional Paper 64, The Centre for Independent Studies, St Leonards, NSW.

Klepper, Steven. (1997) "Industry Life Cycles", Industrial and Corporate Change, Vol.6, No.1, pp145-181.

Low, M.B. & Abrahamson, E. (1997) "Movements, bandwagons and clones: Industry evolution and the entrepreneurial process", Journal of Business Venturing, Vol.12, No.6, pp435-457.

Sarason. Y. Dean, T. and Dillard, J. (2005) "Entrepreneurship as the nexus of individual and opportunity: A structuration view", Journal of Business Venturing, Vol.21, No.3, pp286-305.

Sarasvathy, S.D., Dew, N., Velamuri, R. & Venkataraman, S. (2002) "Three views of entrepreneurial opportunity". In D.N. Audretsch & Z.J. Acs (eds.) Handbook of entrepreneurship research, Kluwer Academic, Boston, Mass and London.

Schumpeter, J. (1934) The Theory of Economic Development, Harvard University Press, Cambridge.M.A.

Shane, S. (2000) "Prior knowledge and the discovery of entrepreneurial opportunities", Organizational Science, Vol.11, No.4, pp448-469.

Shane, S. (2003). A general theory of entrepreneurship: The individual-opportunity nexus, Edward Elgar, Cheltenham, UK.

Shane, S. & Eckhardt, J. (2003). The individual-opportunity nexus. In J. Zoltan, Z. Arcs, & D. Audretsch (eds.) Handbook of entrepreneurship research. Kluwer Academic Publishers, Dordrecht.

Shane, S. & Venkataraman, S. (2000) "The promise of entrepreneurship as a field of research", Academy of Management Review, Vol.25, No.1, pp217-226.

Timmons, J.A. (1994). Opportunity recognition: the search for higher potential ventures. In W.D. Bygrave (ed.), The portable MBA in entrepreneurship, Wiley, New York.

Timmons, J. (1999) New venture creation: Entrepreneurship for the 21st century, Irwin McGraw Hill, Burr Ridge, Ill.

Yoffie, D. B. Bergenstein, Sigrid (1985) "Creating political advantage: The rise of the corporate political entrepreneur", California Management Review, Vol.28, No.1, pp124-139

Weidenbaum, M.L. (1980) "Public Policy: No longer a spectator sport for business", The Journal of Business Strategy, Vol.1, No.1, pp46-53.

The Process of Social Innovation: Multi-Stakeholders Perspective

Kanji Tanimoto

Hitotsubashi University, Tokyo, Japan

First published in The Proceedings of ECIE 2010.

Editorial Commentary

Social entrepreneurship and social innovation refer to entrepreneurial and developmental activities undertaken by social entrepreneurs – excluding public institutions on the macro level - to solve problems of social integration, socially dysfunctional behavior and socio-economic development (Bright & Lindsey, 2010; Chell, E; Nicolopoulou, K; Karataş-Özkan, M., 2010). The concept is relatively new and the field is still in its infancy. But in search of a working definition, researchers have underscored that attributes and talents of social and conventional entrepreneurs are similar – innovativeness, tenacity, resilience – and the difference resides in the motivation and purpose: social value creation against financial need (Roberts & Woods, 2005).

In this emerging field, Tanimoto's work offers a significant contribution by describing the *process* of social innovation, i.e. social innovative business activities undertaken by social entrepreneurs with creation and diffusion of social value, and the essential relationship with stakeholders, as the outcome. The notion of "open innovation" underlies the development of what is called a "social innovation cluster" with emphasis on collaborative relationship, flexibility and open access. Well structured, well documented, it constitutes a solid reference for this stream of research which is enhanced by the clarity of the presentation.

References

Bright David S, Godwin Lindsey N.2010. Encouraging Social Innovation in Global Organizations: Integrating Planned and Emergent Approaches. Journal of Asia - Pacific Business. Binghamton: Jul 2010. Vol. 11, Iss. 3;

Chell, Elizabeth; Nicolopoulou, Katerina; Karataş-Özkan, Mine, 2010..Social entrepreneurship and enterprise: International and innovation perspectives. Entrepreneurship & Regional Development, Vol. 22 Issue 6, p485-493.

Roberts, Dave; Woods, Christine , 2005. Changing the world on a shoestring: The concept of social entrepreneurship. University of Auckland Business Review, Autumn2005, Vol. 7 Issue 1, p45-51, 7p

Abstract: This paper focuses on the process of how social innovation is created. Recent years have seen the increasing emergence of social entrepreneurs, who are expected, as social innovators, to tackle social problems and change society. There have been many case studies on and research about social enterprises and entrepreneurs. However, few studies have been made about social innovation in comparison to the vast volume of research into business innovation. A numbers of studies focused on a single charismatic social entrepreneur concentrate on the description of his/her success story, with little attention paid to the process of social innovation. Most social innovation is not created by a single entrepreneur. There is a need for research into how social entrepreneurs encounter collaborative stakeholders, and how they bring about social innovation through unique ideas. Business innovation studies have shifted focus from the closed process of creating innovation within an organization to the open process: now, many studies argue that customers and users play an important role in creating innovation. In the case of social innovation, it is created not only by entrepreneurs and producers alone, but by various related stakeholders as well as customers and users. The social entrepreneur identifies social problems, gets ideas and resources, and creates social innovation in collaboration with related stakeholders. This paper tries to clarify this process through a case study of the Hokkaido Green Fund (HGF), an environmental NGO. HGF has introduced the first community wind energy business to Japan. This system, built by the local community, is not the primary innovation but a secondary/derivative innovation. It is not easy to transplant social innovation created in another area or country to a new location. In the process of introducing and creating the system, entrepreneur has been supported by and collaborated with related stakeholders, and has had an impact on the social system. This paper tries to present a new perspective for the analysis of the social innovation process, from the viewpoint of multi-stakeholders.

Keywords: social innovation, multi-stakeholders, derivative innovation, collaboration

1. Introduction

Social entrepreneurs have been emerging and expected as new social innovators. They have tackled various social problems such as welfare, community development, environment, and cooperation with developing countries through business activities rather than volunteer activities. Social enterprises are expected to provide new innovative business models in social fields, able to respond to a variety of social needs in the local and global communities, to which conventional schemes are not able to respond.

The roles and potential of social entrepreneurs have been spotlighted by media and academia alike recently, and studies on social entrepreneurship have been increasing. There are already many case studies and theoretical studies on social enterprises and entrepreneurs. However, in comparison with the vast volume of studies on business innovation, there is a dearth of academic research on how social innovation is created. The purpose of this paper is to clarify the process of social innovation.

Although some characteristics of social innovation are similar to business innovation, others are rather different. It is true that some of the concepts and frameworks found in studies on business innovation are adaptable to social innovation. However, social innovation displays many unique characteristics, primarily because social enterprises have a mission with a double bottom line: to achieve social performance as well as economic performance.

Now, we define the concept of social enterprise and social entrepreneur. The Office of the Third Sector of UK Government defines social enterprise as follow: a social enterprise is a business with primarily social objectives whose surpluses are principally reinvested for the purpose of the business or in the community, rather than being driven by the need to maximise profit for shareholders and owners. The same office defines a social entrepreneur as a person driven by a desire to change society. This definition focuses on the social dimension of a social entrepreneur. Dees and Anderson (2006) insist that a social entrepreneur plays the role of a change agent in the social sector by adopting a mission to create and sustain social venture. They concentrate on those social entrepreneurs who carry out innovations that blend methods from the worlds of business and of philanthropy to create social value. This definition focuses on the innovative dimension of a social entrepreneur.

The concept and the corporate form of social enterprises vary from country to country (Kerlin 2006), depending on social context and the maturity of the civil society concerned. The following are integral and indispensable factors of social enterprise: 1) Social mission: to have a clear mission of tackling social challenge(s) and facilitating social change. Social enterprise can operate its business only by receiving support from stakeholders. 2) Social business: to create a new comprehensive business model to realize its social mission. It is difficult for social enterprises to venture into the social field with no chance to make profit, but their purpose remains focused on the creation of new social value rather than the maximisation of profit for shareholders and owners. 3) Social innovation: to develop new social goods and services, and unique systems to address social problems. It is also important that the social business prompts the realization of new social values. It is no easy task to connect the realization of social mission and the performance of profitable business. It is social innovation itself that connects these two factors.

2. Social innovation

2.1. Social innovation theory

Aiming to change society and making business work well are not actions that are linked automatically. Social entrepreneurs who are able to connect both and to develop unique activities in the process are creating innovation. Social entrepreneurs are not necessarily required to create new technologies, materials or product innovation, but to develop new schemes and unique business models.

Innovation is generally defined as that which introduces something new, makes changes in anything established. Innovation of economic activity means innovation which brings economic effects. Drucker (1985) points out that entrepreneurs create something new and something different and change or transmute values. This idea also applies to social entrepreneurs. Muglan et al. (2007a) think of social innovation as the development and implementation of new ideas (products, services and models) to meet social needs. This paper defines social innovation as innovation which creates new social values through businesses which tackle social problems with a view to their resolution.

Studies on social innovation have been popular these past few years. In comparison to the vast volume of research on business innovation, however, there is a remarkable dearth of academic research that looks at how social innovation is created and analyses the process of social innovation. Mulgan et al. (2007a) have argued that "the competitive pressures that drive innovation in commercial markets are blunted or absent in the social field", but the situation has been changing rapidly. Growing global attention and the boom on social entrepreneurs have intensified the research being carried out on social entrepreneurs and their innovative activities.

There are a variety of discussions on social innovation, ranging from innovative political and welfare systems through macro-institutional change (Hämäläinen et al. 2007) to innovative business models by social entrepreneurs. Drucker (1985) argues that social innovation includes not only technology but also frameworks of insurance and healthcare which have a huge impact on society. He analyses innovation strategies of public-service institutions (government agencies, universities, hospitals, non-profit organisations in the community) as well as business and new ventures. He explains the main features and policies of social innovation by public-service institutions, but does not analyse how the social innovation itself is created. This paper focuses on innovative business activities by social entrepreneurs, not on innovation by public institutions on the macro level. There are some studies which discuss new movements, led by entrepreneurs, which are tackling social problems. For example, Westley et al. (2006) analyse the innovative approaches of various players, including government, NPOs, volunteer groups, financial groups and business corporations, regarding social subjects including HIV/AIDS in the community, crime prevention, and support for the disabled. Mulgan et al. (2007a) examine the characteristics of the different approaches shown by various players, including NPOs, government, markets, movements, academia and social businesses, regarding fair trade, hospices, correspondence courses, open universities and

Wikipedia. These studies deal with political and social issues at the community level and analyse the structural mechanism of reform and the meaning of social innovation; they are not, however, necessarily focused on business schemes.

Studies focused on the social innovation of social enterprises are increasingly common. Dees (1998) defines social enterprise as being located in the centre of two points on a linear scale: the purely charitable and the purely commercial. Social entrepreneurs, who can be called change agents, seek out opportunities to improve society, to create new social values. They consider social innovation as their fundamental resources; new and better ways of serving their social mission (Dees et al., 2001). They regard social entrepreneurs as promoting innovation which matches their social business and philanthropic activities in order to create social value. Dees et al. (2001) mainly research the strategic management of social innovation, however, and not the process of social innovation.

Discussion on how to make social innovation work has been led by a study by Mulgan et al. (2007b). They point out that "the successful growth of social innovations depends on effective demand and effective supply coming together", and that "innovations often begin with simple ideas and insights, which may ultimately originate from many different sources including social entrepreneurs, bureaucrats, frontline staff, service users, observers or volunteers". Moreover they suggest that the "diffusion of an idea" is the key point in developing more effective social innovation. Social enterprises need "effective strategies" (choices about supporters and organisational form) and "learning and adaptation". They insist that the key issue is how to connect 'pull factors' coming from government and the community to 'push factors' coming from those who have ideas. As they put it, "the combination of 'effective supply' and 'effective demand' results in innovations that simultaneously achieve social impact and prove to be financially sustainable". They explain the mechanism of social innovation from the viewpoint of demand and supply, but do not clarify the process of how social entrepreneurs create social innovation.

Westley et al. (2006) assert that the idea of complexity explains the process of how social innovation is created within the interactions of various movements and how it changes society. They suggest that "relationship is a key to understanding and engaging with the complex dynamics of social innovation" and that "for social innovation to succeed, everyone involved plays a role. As sift, everyone--funders, policy makers, social innovators, volunteers, evaluators--is affected. It is what happens between people, organisations, communities and parts of systems that matters 'in the between' of relationships". This idea, which considers social innovation as being in a dynamic relationship with stakeholders, is thought-provoking for our research. However, their perspective of complexity remains nothing more than an idea, and they do not go on to explain the mechanism and process of social innovation.

Christensen et al. (2006) refer to disruptive innovation for social change as "catalytic innovation": "What's required is expanded support for organisations that are approaching social-sector problems in a fundamentally new way and creating scalable, sustainable, systems-changing solutions." Here, innovation presents a new possibility to under-served people whose needs have not been met in areas with insufficient social services. They pick up some cases, such as low-cost medical insurance and affordable education programs, e-learning at secondary schools, community colleges, and micro-lending systems, made available to people who otherwise would have limited or no access to educational opportunities. However, they also don't explain the processes behind the birth and development of social innovation.

As we've seen, social innovation is already being considered from a variety of perspectives. The specific purpose of this paper, however, is to clarify the processes by which social innovation is created.

2.2. The creation of social innovation

This section considers the basic framework of how to analyse the process of social innovation. The primary questions here are where, and by whom, is social innovation created?

Business innovation studies have focused on whether innovation is created within the organisation (research and development division, project team) or outside the organisation (user/customer, collaboration with other actors), as well as whether innovation is created in a closed or an open process (Chesbrough 2003).

In researching the process of social innovation as it addresses social problems, studies on user-led innovation and user/producer co-created innovation are useful, rather than those which concentrate on producer-led innovation. For example, Ogawa (2000) explains user innovation in terms of a "Sticky Information Hypothesis" (where the costs of sticky innovation-related information would have an impact on the locus of innovation) inspired by von Hippel (1994). Ogawa (2006) also explains that innovation-generating collaborative activities between the producer and the user are competitive. Von Hippel (2005) points out that the user's ability and environments to generate innovation are developed, not by the producers who are the providers of products and services in various areas. Prahalad et al. (2004) focus on the process of value creation by consumer-company interactions. With the global spread of the internet, consumers now have access to a large volume of information and are able to create online communities and new values beyond conventional geographic and social boundaries. Consumers become committed to interactions and co-creations with firms.

Up until now, studies on social enterprise have primarily concentrated on case studies, entrepreneur history studies and management studies. Few have focused on how social innovation is created and how it changes society, or on the discovery of the processes involved in social change. In general, studies have focused on a single charismatic entrepreneur, describing his or her success story. Not all social innovation, however, is produced by a single entrepreneur.

How are social entrepreneurs finding out about social problems, creating business schemes with unique ideas and resources, and diffusing them? In many cases, social innovation is created not only with users and customers, but in collaboration with various stakeholders. As the case of the Hokkaido Green Fund, considered in the next section, social innovation is created through an open relationship with stakeholders and a collaborative process with them. The focus of this research is to analyse the relationship between the entrepreneur(s) and stakeholders, and the dynamic process of creating social innovation.

This point relates to the suggestion of Matsushima and Takahashi (2007) that a new perspective is necessary to clarify the dynamic process in which the institutional entrepreneur comes to have relational rules with various actors. This is closely linked to the idea that the "paradox of embedded agency" should be deciphered; which explains that, as they come to have the cause and the opportunity, entrepreneurs try to gain resources in order to change a certain system, while being embedded in that system. The

key point is to understand the process of how entrepreneurs create innovation and introduce possibilities for social change from their relationship
with various stakeholders.

3. Social innovation in the Hokkaido Green Fund

When a social entrepreneur recognizes a social problem and starts a new
social business targeting it, he or she thereby creates a relationship with
various stakeholders and collaborates with them. This section reviews this
process in the context of a case in Japan: the first community wind energy
business, built by the Hokkaido Green Fund (HGF) in collaboration with
local people and organisations. The following content is based on several
interviews with HGF members and HGF internal documents.

3.1. The identification of social issues

The Hokkaido Green Fund, an environmental NGO, was established in Sapporo City, Japan, in July 1999 (Chairperson: Sakae Sugiyama; Director-
General: Toru Suzuki). Its purpose is to enable citizen to play a positive role
in creating energy innovation themselves, without being limited to con-
ducting an anti-nuclear power plant movement. The starting point, however, can be traced back to an anti-nuclear movement, the "Good-Bye Nuclear Group", started in 1988, of which Suzuki used to be the group leader.
It consisted of members mainly of the Seikatsu-Club Consumer's Cooperative Union Hokkaido, of which Sugiyama used to be Chairperson from 1986
to 1998. The movement was triggered when Sugiyama and the members
of the Cooperative Union encountered radioactively-contaminated vege-
tables, the contamination having been caused by the nuclear meltdown at
the power station in Chernobyl. And the members learned the limitations
of their efforts when they were not able to prevent the construction of
nuclear power plants in Tomari, Hokkaido, as planned by the Hokkaido
Electric Power Co. After this set-back, they began to refer to alternative
business-styled movements in Europe, and remodelled the basic strategies
of their activity on 'a style of movement incorporating practical business,
and a business style incorporating a sustainable movement'. Then, they
stepped up efforts to establish a community wind energy business.

At that time, wind-powered electricity businesses established through citi-
zen-led initiatives were already popular in Europe. The activities of HGF did
not represent the first innovation in this field, although they were un-

precedented in Japan. Back then, Japan's policy on electric power was rigid, and electric power companies dominated the electricity market in each region. Although free access to the electricity market was granted in 1995, the entry of citizens into that market was not seen until the HGF was established. They were passionate about creating a new model to change the rigid energy policy. Redlich (1951) describes original and unprecedented innovation as "primary innovation", and innovation which has been produced elsewhere but is introduced into a new area as "derivative innovation". It is not easy to transplant a foreign model into one's own country and make it succeed, because of differences in social structure, resources and value. Imitation of an established model alone will rarely lead to the successful introduction of innovation. In order to realise derivative innovation, various skills and efforts are needed to adapt the innovation appropriately to each country.

3.2. Collaboration with stakeholders

While looking at the possibility of generating electricity through natural energy, before establishing the HGF, Sugiyama and Suzuki learned about the green electricity tariff scheme, which has proved popular in the U.S., from Prof. Koichi Hasegawa of Tohoku University. They also found out about several cases of wind energy businesses owned by local communities in Denmark from Tetsuya Iida, Director of the Institute for Sustainable Energy Policies. This information helped them to form more concrete ideas, leading them to consider the possibility of running a wind energy business in Japan with reference to preceding cases. First, they introduced a green electricity tariff scheme in cooperation with the Hokkaido Electric Power Co., researched some well-developed cases of community wind energy projects in Europe, and then began to set up a business plan for installing wind turbines and selling the electricity generated by civic hand.

By the end of 1999, Hokkaido Electric Power Co. announced that they would be purchasing natural energy for the remainder of the period ending March 2001. The HGF Board of Directors was forced to make a quick decision committing to starting the business, saying "we cannot tell when the next opportunity will come". Fundraising presented a severe problem for the HGF, since approximately 200 million yen (approximately 2 million dollar) is needed to purchase and install a single wind turbine. In the first instance, the collection of such an amount of donations would have been entirely impossible for them. A further problem was that of the institutional restriction applied to non-profit organisations, not being able to receive investment from the market. After hard negotiations with Hokkaido banks, they finally received a response from North Pacific Bank: they would lend the HGF 140 million yen provided it established a new business corporation, clarified its accountability and raised 60 million yen on its own.

HGF then decided to establish a business corporation for fundraising in the market (to develop their business through a combination of two different styles of organisations: a nonprofit organisation and a company corporation).They met Hiroyuki Kawai, an attorney and the founder of Sakura Kyodo Law Offices, who had an interest in this business and provided the HGF with advice and legal support in establishing the corporation (Kawai later became the auditor of the Natural Energy Community Fund Ltd., which was established later by the HGF). In February 2001, Hokkaido Citizens' Wind Power Co. Ltd was officially launched, with 14 stockholders (13 individuals, 1 corporation=HGF), and Toru Suzuki, who had been a leader of the business plan, was appointed President. The company has a council system with three members, President Toru Suzuki, Vice-President Sakae Sugiyama, and Yotaro Kashiwa (the representative of Anti-Nuclear Citizens' Group), having right of representation. Anonymous union investment to the company's business was adopted to raise capital. Starting in December 2000, by asking for an investment of 500,000 yen per contribution, the HGF was able to collect investment totaling around 100 million yen in the first month alone. In Sapporo City, a voluntarily project, entitled the "Community Wind Power Supporter Group", was also organised. This group asked for investments of minimum 50,000 yen per contribution, and established a system whereby up to 10 smaller units of contribution could be combined to make up the minimum total, thereby making it easier to contribute with fewer funds. They succeeded in collecting 5 million yen in the three months from May to July. The movement was accelerated by such support networks, and was able to gain public acceptance and support from the community in a short period. By September 2001, the total amount raised stood at 156.5 million yen collected from 212 persons and 17 corporations, and the total business fund totalled 166.5 million yen, with the extra 10 million yen coming from the green energy tariff system and donations from labour unions. The HGF also received a loan of approximately 70 million yen from North Pacific Bank, and Japan's first community-run wind turbine was finally installed in Hamatonbetsu-cho, Hokkaido, in September 2001 (Generator: 990kw, annual output: 2.6 million kw).

This business was supported by many people and organisations. In starting HFG, they gained the full cooperation of Hokkaido NPO Centre. Toshio Hori, at that time Director of the Electric Business Division of Tomen Power Japan Ltd, cooperated positively with HGF including making suggestions for appropriate site locations, research and technical advice on the installation of wind turbines. Akira Otani, another employee of Tomen Power Japan, later joined the HGF and played a core role in the development and man-agement of the business. Figure 1 charts the relationships amongst the related stakeholders.

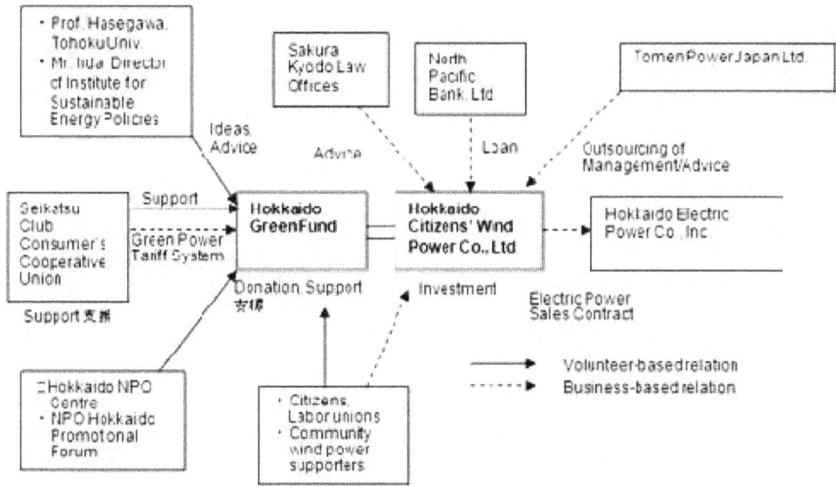

Figure 1: Hokkaido Green Fund and its stakeholders

There were two major factors that explain why cooperative activities and investment fundraising were successful in this case: 1) the public's aware-ness of and negative attitude towards nuclear power, and 2) the public's expectations for the first community wind energy business in Japan. The success of this community wind energy business owes much to its many sponsors and supporters. In this project, stakeholders with various pur-poses have come together to collaborate on the basis of value sharing through common experiences. It appears to be a major factor of social in-novation that stakeholders share a common experience, here taking part in establishing the community wind energy business. Social innovation with a correlation of stakeholder interest can change social value. A questionnaire, carried out on investors in community wind energy by Iida et al. (2003), shows that many of the respondents were passionate about wanting to be part of solving local environmental/energy issues. One resident of the village where the wind turbine was installed stated: "we are now proud of the wind, which until now has only been a distraction to us". After the success of the first installation, a second wind turbine was constructed based on the experience and systems of the first one. This in turn encouraged people in other areas to construct more community wind energy plants, indicating that the innovation has begun to diffuse to other areas. Later, in February 2003, the HGF established the Natural Energy Community Fund Ltd. to build a new network of development and support for community wind power. The HGF has entered the second stage of its business, and this scheme of running community-based wind power projects is expected to develop throughout other areas in Japan.

4. Concluding remarks

The HGF case shows that there is a correlated relationship between related stakeholders who share experiences in the process of creating social innovation in a local area. Inspired by a certain social mission, social entrepreneurs source diverse ideas and resources from stakeholders, and then collaborate to start a business. In clarifying this relationship, we find that there exists a structure which can be deemed a "Social Innovation Cluster" in the area. This cluster is defined as an organisational accumulation that includes social enterprises, support organisations, funding agencies, universities and research institutions. By building cooperative relationships in a cluster, new social businesses are born and they generate and provide innovative social solutions and social values. The Social Innovation Cluster has similarities to the Industrial Cluster, but it also has its own unique characteristics; it is more flexible and more community-rooted. The basic characteristics are cross-section, interaction with its community, and open access.

Social enterprises affect stakeholders through their business activities, at the same time, it is impossible for them to exist without being accepted by those stakeholders. Stakeholders recognise and come to learn about social issues from their business activities. New social value can be realised

through purchasing and supporting of goods and services provided by social enterprises with a social message. People who encounter the message
ill experience increased awareness of and concern for social issues. They
will come to recognise significant social problems, share their values and
then become involved with community and social problem-based activities. In other words, social value is realised through peoples' experience
and practice. In the case of the HGF, people who have been inspired by the
HGF's activities have experienced making an investment in social business,
witnessing the installation process of the wind turbine, and recognising its
social impact via reportage by the mass media. Furthermore, their environmental consciousness has been gradually changing throughout these
experiences; they have developed an interest in other environmental issues, or altered their behaviour to save more energy, or come to participate in social programs on environment.

This can be described as similar to the style of "experience innovation"
proposed by Prahalad *et al.* (2004). Figure 2 describes this process of social
innovation.

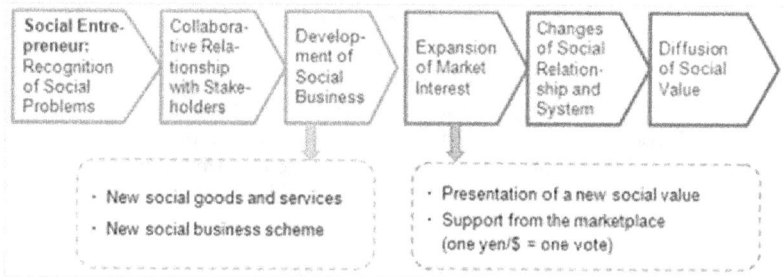

Figure 2: The process of social innovation

References

Chesbrough, H. (2003) Open Innovation: The New Imperative for Creating and Profiting from Technology, Harvard Business School Press, Boston.

Christensen, C. M., Baumann, H., Ruggles, R. and Sadtler, T. M. (2006) "Disruptive
Innovation for Social Change", Harvard Business Review, Vol.84, No.12, pp.94-
101.

Dees, J. G. (1998) "Enterprising Nonprofits", Harvard Business Review, Vol.76, No.1,
pp.55-66.

Dees, J. G., Emerson J., and Economy. P. (2001) Enterprising Nonprofits, John Wiley & Sons, Inc., New York.

Dees, J.G. and Anderson, B.B. (2006) "Framing a Theory of Social Entrepreneurship: Building on Two Schools of Practice and Thought", Research on Social Entrepreneurship, ARNOVA Occasional Paper Series1-3, pp.39-66.

Ducker, P. F. (1985) Innovation and Entrepreneurship, Harper & Row Publishers, Inc., New York.

Hämäläinen, T. J. and Heiskala R. (eds.) (2007) Social Innovations, Institutional Change and Economic Performance: Making Sense of Structural Adjustment Processes in Industrial Sectors, Regions and Societies, Edward Elgar Publishing Ltd., Northampton.

Iida, T., Maruyama, K., Kasuya, I., Suzuki, T., and Hasegawa, K. (2003) "Shiminshutai gata no Energy Seisaku ni kansuru Kenkyu [Research on the Energy Policy by Citizen's Initiative]", Research Institute for Consumer Affairs, [online], http://www.ge-aomori.or.jp/activity/coop_report.pdf

Kerlin, J.A. (2006) "Social Enterprise in the United States and Abroad: Learning From Our differences", Research on Social Entrepreneurship, ARNOVA Occasional Paper Series1-3, pp.105-126.

Matsushima, N. and Takahasi, T. (2007) "SeidotekiKigyoka no GainenKitei: Umekomareta Agency no Paradox ni taisuru Rironteki Kousatsu [Concept of Institutional Entrepreneur: theoritical study on the paradox of embeded-agency]", Discussion Paper, Vol.48, Kobe University, Kobe.

Mulgan, G., Ali, R., Halkett, R. and Sanders, B. (2007a) "In and Out of Sync: The Challenge of Growing Social Innovations" NESTA, London.

Mulgan, G., Tucker, S., Ali, R. and Sanders, B. (2007b) "Social Innovation: What It Is, Why It Matters and How It can be Accelerated", Skoll Centre Oxford Said Business School, Oxford.

Ogawa, S. (2000) Innovation no Hassei Riron [The Locus of Innovation], Chikura-shobo, Tokyo.

Ogawa, S. (2006) Kyôsoteki Kyôso Riron [Competitive Co-Creation], Hakuto-shobo, Tokyo.

Prahalad, C.K. and Ramaswamy, V. (2004) The Future of Competition: Co-Creating Unique Value With Customers, Harvard Business School Press, Boston.

Redlich, F. (1951) "Innovation in Business: A Systematic Presentation", American Journal of Economics and Sociology, Vol.10, No.3, pp.285-291.

Rogers, E. M. (2003) Diffusion of Innovation, Fifth Edition, Free Press, New York.

Von Hippel, E. (1994) "Sticky Information and the Locus of Problem Solving: Implications for Innovation", Management Science, Vol.40, No.4, pp.429-439.

Von Hippel, E. (2005) Democratizing Innovation. The MIT Press, Cambridge.

Westley, F., Zimmerman, B. and Patton, M. (2006) Getting to Maybe: How the World is Changed, Random House of Canada Ltd., Toronto.

Towards Collaborative Open Innovation Communities

Maria Antikainen

VTT Technical Research Centre of Finland, Tampere, Finland
Tampere University of Technology (TUT), Finland

First published in The Proceedings of ECIE 2010.

Editorial Commentary

This volume ends by addressing the topic of Open Innovation and Open Innovation Communities. Since Chesbrough coined the term "Open Innovation" in his seminal work in 2003, innovation openness is becoming the prevailing innovation model inducing different approaches: from (collaborative) open innovation related to innovation by multiple, dispersed actors related to a public good, to the acquisition of new ideas, patents, products, etc. from outside its boundaries (Chesbrough 2003; Baldwin & von Hippel, 2010). Web2.0 technologies have fostered the capabilities of innovation openness by allowing large, dispersed groups of people to collaborate on innovative tasks, based on the power of collective intelligence (Howe, 2008).

From a Human Resource Management perspective, this exploratory study by Antikainen investigates the collaboration in online open innovation communities by addressing the factors that foster users' motivation and the main components of a rewarding strategy. Motivation is encouraged by the quality of, and interest in the stated objectives. Equity and democracy appear to be the main drivers of a successful rewarding strategy. Attention is also drawn to collaboration methods and tools and their facilitating / inhibiting effects. The contribution serves both the academic and practitioner literatures by providing guidelines for OI community management.

References

Baldwin Carliss, von Hippel Eric, 2010. Modeling a Paradigm Shift:: From Producer Innovation to User and Open Collaborative Innovation, MIT Sloan School of Management Working Paper # 4764-09; Harvard Business School Finance Working Paper No. 10-038

Chesbrough, Henri, 2003. The Era of Open Innovation, MIT Sloan Management Review, Spring 2003, p 35-42

Howe, J. 2008. Crowdsourcing: Why the Power of the Crowd is Driving the Future of Business, RandomHouse, New York,

Abstract: Open innovation (OI) communities have changed our conceptions of how innovation can and should be managed and have prompted calls for new theories of innovation. OI communities with customers can act as a source for learning and producing external ideas or even solutions to companies. As earlier studies indicate that collective problem solving improves the quality of ideas, motivating and supporting collaboration in online OI communities is important. This explorative study explores collaboration in online OI communities by answering two questions. The first question considers users' motivations to collaborate in OI communities, while the second one explores how rewarding can be used to motivate collaboration in OI communities. The study consists of three cases: CrowdSpirit, FellowForce and Owela. The preliminary results are based on the data gathered by interviewing maintainers of the communities and by a questionnaire to the community members. According to the results, the users were motivated to collaborate by interesting objectives and the concept of the community, gaining new viewpoints from other users, obtaining better products and receiving rewards. The results also indicate that the lack of proper tools inhibits collaboration in OI communities. Furthermore, an OI community's rewarding strategy should be transparent and logical. Rewarding should be based on the efforts and quality of the work rather than on giving rewards based on the quantity of ideas or lotteries. The system should be flexible so that rewards vary in different situations and phases of the work. The equity and democracy of the rewarding system are important factors for OI community users. Additionally, customisability of the rewarding strategy ensures that users can influence, on some level, the nature of the rewards they receive, and the rewards will therefore be more valuable to everyone. This explorative study is one of the first studies of collaboration in online OI communities. In addition to serving academia, the study provides practical knowledge on how to reward and motivate groups of members on the web to companies and the growing number of OI intermediaries building or planning to build innovation communities.

Keywords: online communities, open innovation, intermediaries, rewarding, collaboration, monetary, non-monetary, tangible, intangible, recognition, motivation, case study

1. Introduction

1.1. Background of the study

OI paradigm assumes that firms can and should use external ideas as well as internal ideas, and internal and external paths to market, as the firms look to advance their technology (Chesbrough, 2003, p xxiv). OI specifically with customers, provides interesting possibilities for companies to improve their innovation processes. Online OI communities with customers can serve as a source for learning and producing external ideas or even solutions for companies (Jeppesen et al. 2006; Chesbrough 2006). To integrate customers into innovation processes in online OI communities, companies need methods, tools, platforms and resources as well as different types of services provided by external companies. In order to do this, companies can: 1) build their own OI communities, 2) use existing online communities related to their products and services, such as brand communities, 3) look for hobbyist communities, or 4) use existing communities on the web that act as intermediaries in this field (Chesbrough 2003; von Hippel 2005; Chesbrough 2006a). The number of online innovation market places and innovation intermediaries acting between innovators and companies (or 'solvers' and 'seekers') has recently grown rapidly. As earlier studies indicate that collective problem solving improves the quality of ideas (e.g., Hargadon and Beckhy 2006; Thrift 2006), it is important to motivate and support collaboration in online OI communities. It is therefore relevant to study these issues in order to serve companies managing their own communities, innovation intermediaries offering online OI communities as a service, and academia.

1.2. Purpose and methodology

This explorative research studies collaboration in online OI communities by addressing two questions. The first question considers users' motivations to collaborate in OI communities. The second question concentrates on how rewarding can be used to motivate collaboration in OI communities. The study consists of three cases: CrowdSpirit, FellowForce and Owela. Our research team gathered the data by interviewing maintainers of the communities and by a questionnaire to the community members. Semi-structured interviews with the maintainers were conducted by phone and recorded. The recordings were transcribed as notes afterwards. The data were collected March-April 2008. The interviews each lasted approximately one hour and covered questions related to the members' motivation factors, existing collaboration tools and methods, and future plans to support collaboration. Also the communities' rewarding models were discussed. In addition to the interviews, a web survey was conducted covering themes related to collaboration and aspects related to rewarding and motivation. One hundred of FellowForce's top members and two hundred of CrowdSpirit's most active members were contacted by email and asked to participate in the survey. This plan of action was chosen due to the wish of the maintainers instead of using a link on the web site. However, in the Owela's case we used the link on the web site and in aim to get more responses and the survey was also marketed in Owela's newsletter, which was sent to its members.

2. Literature review

2.1. Users' motivations to collaborate in online OI communities

In this study the focus is on online OI communities where users participate in organisations' innovation processes at some level. To encourage collaboration online OI communities offer set of tools as well as utilise different methods, such as rewarding. The first step to knowing why users collaborate in online communities is to understand their motivations to participate in and contribute to online communities. Studies into why people visit, join, participate in and contribute to different kinds of online communities have been carried out from varying perspectives. Prior literature has divided human motivation into intrinsic and extrinsic reasons (Deci and Ryan 1985; Amabile 1996; Ryan and Deci 2000). Motzek (2007) also stressed the impact of social motives in a person's code of conduct and therefore added a third category for social motives.

According to Deci and Ryan (1985), intrinsic motivation refers to situations in which a person does something because it is inherently interesting or pleasant to do so. In contrast, extrinsic motives refer to the consequences of a certain activity, perceiving the task itself as a means to an end (Amabile 1993). As online communities are based on the idea that members share some common interest and purpose, social motives play an important role in sustainable online communities. Social motives can also be seen as essential when enhancing collaboration between members. Wasko and Faraj (2000) stated that people do not use the forum to socialise or develop personal relationships. According to their study, giving back to the community in return for help received was by far the most cited reason for people's participation. Table 1 divides the closely related motivations into three main categories: intrinsic, extrinsic and social motives. Most categories include motivation factors identified in different kinds of communities, such as OSS, firm-hosted communities and OI communities (Antikainen & Väätäjä 2010).

Prior research into users' motivations to collaborate proposes that collective work with others is seen as enriching, fun, productive, efficient and even the best way to trigger creative innovation. Furthermore, it suggests that collaboration should be sought in order to get the most out of people's creativeness. The research admits, however, that it is demanding to support collaboration in an online environment as it is already challenging to create collaboration between strangers in face-to-face situations (Antikainen et al. 2010).

2.2. Motivation and its relationship to rewarding

Rewarding can be divided into monetary (tangible) rewarding and non-monetary (intangible) rewarding (also called recognition). Monetary rewards include money, pay checks, fees, trophies and awards. Non-monetary rewarding may be the member's name in honour-roll lists or top ten lists, granting privileges and public recognition. (Antikainen & Väätäjä 2010)

Studies in the field of psychology suggest that expected monetary rewards tend to reduce intrinsic motivation, whereas praise and other positive verbal feedback tend to increase it (Deci 1975, and Lepper et al. (1973). According to Reeve's (2005) studies into rewarding and its relationship to intrinsic motivation, extrinsic rewards for intrinsically interesting activity have a negative effect on future intrinsic motivation. Several studies have implied that the expectancy and tangibility of the reward reduce the intrinsic motivation when a person expects a reward for a completed task. No widely accepted theories on the relationship between motivation and rewarding currently exist however (Lindenberg, 2001). The simplicity of the

theories on motivations and study setups presumably cause some misin-
terpretations, as in real-life several motivations may exist concurrently.

Table 1: Users' motives for participating in online communities (Antikainen
& Väätäjä 2010)

Intrinsic motives	Ideology	Lakhani and Wolf 2005; Stewart and Gosain 2006; Nov 2007
	Enjoyment, fun, recreation	Raymond 2001; Torvalds & Diamond 2001; von Hippel and von Krogh 2003; Osterloh et al. 2004; Ridings and Gefen 2004; Lakhani and Wolf 2005; Nov 2007
	Intellectual challenges, stimulation, interesting objectives	Ridings and Gefen 2004; Lakhani and Wolf 2005
	Learning, improving skills and knowledge exchange	Wasko and Faraj 2000; Hars and Ou 2002; Wiertz and Ruyter 2007; Antikainen 2007; Gruen et al., 2005
Extrinsic motives	Firm recognition	Jeppesen and Frederiksen 2006
	Reputation, enhancement of professional status	Bagozzi and Dholakia 2002; Lernel and Tirole 2002; Lakhani and Wolf 2005
	Sense of efficacy	Constant et al. 1994; Bandura 2005
	User need, influencing the development process	Hars and Ou 2002; von Hippel 2005; Lakhani and Wolf 2005;
	Rewards	Antikainen and Väätäjä 2008; Wasko and Faraj 2000; Lakhani and Wolf 2005; Harper et al. 2007; Kittur et al. 2008
Social motives	Altruism, reciprocity, care for the community	Kollock 1999; Wasko and Faraj 2000; Zeityln 2003; Nov 2007; Wiertz and Ruyter 2007
	Friendships, "hanging out together"	Rheingold 1993; Hagel and Armstrong 1997; Ridings and Gefen 2004
	Peer recognition	Jeppesen and Frederiksen 2006;

		Lerner and Tirole 2002

The idealised picture of online communities seems to be that the members' contribution is primarily related to intrinsic motivation such as fun, ideology and challenges. Despite some positive results concerning rewarding and motivation (Lakhani and Wolf 2005; Harper et al. 2008), the predominant belief appears to be that no monetary rewards are needed and only non-monetary rewarding or unexpected rewards would be satisfactory to members. More studies are needed in an online community context to explore whether it is in fact a combination of intrinsic and extrinsic motivation and the expectancy to be rewarded for work that is well done for an agreed set of rules. Even fewer studies are available aimed at increasing users' motivations to collaborate by rewarding.

3. Results

3.1. Case communities – brief descriptions and maintainers' interviews

3.1.1. CrowdSpirit

CrowdSpirit (www.CrowdSpirit.com), which originated in France, focuses on electronics design. The founders and maintainers of CrowdSpirit have built toolkits for users to submit their designs and ideas. Similarly, CrowdSpirit includes tools for commenting on and voting for different designs.

CrowdSpirit recently changed its business model. Instead of participating in the development and industrialisation process of the products, CrowdSpirit is now only involved in the design process. In other words, after the design and collection of the project team, the team negotiates directly with the manufacturers. After collecting the project team and drawing up the specifications and a marketing plan, the project team asks for quotations for the development.

In CrowdSpirit, members are used to collaborate with others. They discuss and rate others' ideas and work together in the product design process. To be willing to collaborate, people have to be open, have their own competitive core (more value in this field). According to CrowdSpirit's maintainer, it is a mistake to think that users would collaborate voluntarily.

The maintainer emphasised the importance of rewarding and, more specifically, monetary rewards as compensation for users' work. In the maintainer's opinion, having fun and acquiring new viewpoints from others are the top motivators. Furthermore, the maintainer believes that being in touch and working with companies is motivating for the participants. Collaboration in groups is already a way of working at CrowdSpirit. More collaborative tools such as chat are needed however. The main difficulty is to have people in the community at the same time, as there are members all over the world. To support collaboration, people also need information, profiles and to get to know each other, other than just professionally.

3.1.2. FellowForce

FellowForce (http://www.FellowForce.com) is an innovation marketplace and intermediary that allows companies to submit innovation challenges to solvers. FellowForce originated in the Netherlands and Poland. Solvers provided suggestions (pitches) to a challenge, and the best solvers were rewarded. FellowForce allows solvers to submit their own pitches to companies. Normally, the best pitches matching the challenges are rewarded with money.

Collective creativity is realised in FellowForce's "Innovate Us" functionality. Any registered participant may submit an innovation and view the responses from other users, if this feature is turned on. FellowForce also offers services for companies to launch their own co-creation platforms on their websites.

FellowForce's maintainer stated that members participate because of their curiosity: they just want to try it out. They are also motivated by the possibility of influencing an outcome and sharing ideas with others. FellowForce's maintainer also added that rewarding is a solid part of a sustainable OI community and that it is currently considering ways to enhance collaboration between members with appropriate methods and a rewarding system.

3.1.3. Owela – open web laboratory

Owela (http://owela.vtt.fi) is a participatory web laboratory for designing digital media products and services. Owela was developed at VTT in Finland and aims to connect users to developers and researchers, and to promote OI. Owela offers social media tools for gathering user needs and development ideas and collecting feedback for scenarios and prototypes.

In IdeaTube, users may participate by commenting on the descriptions and visualisations of different situations, needs, ideas, scenarios and proto-types. In TestLab, the prototypes of future products and services can be tested at the beta phase, and the users are expected to give feedback and development ideas. Owela has been used as an innovation platform in re-search projects and studies conducted for companies, such as usability testing of websites.

Owela's maintainer believes that interesting objectives and appropriate tools for participation and collaboration are central factors of user motiva-tion. The maintainer stated that collaboration with others is fun, nourishes creativity and that members learn from each other. It is therefore currently developing tools and methods aimed at enhancing collaboration between members. Appropriate monetary and non-monetary rewarding models are needed to enhance motivation and collaboration.

3.2. The Survey

3.3. Survey respondents

There were 49 responses to our survey. Of the respondents, 45 (91%) were male. The average age of the respondents was 37 years (avg 36.76, std 11.57, min 19, max 64). Almost half of the respondents were members of CrowdSpirit (49%, 24 respondents), 16.3% (8 respondents) of FellowForce, 24.5% of Owela (12 respondents) and 10.2% (5 respondents) of other online OI communities.

3.3.1. Survey responses

Figure 1 shows that 75% of the respondents agreed or strongly agreed that members' mind collaboration was an efficient way to work. Hedonistic factors such as enjoyment and utilitarian factors such as efficiency there-fore rationalise why collaboration is the preferred way of acting in open innovation communities.

All of the communities emphasised the importance of collaboration in the community and were searching for and developing new ways to collabo-rate.

The results in Figure 3 show that rewarding everyone in the group was not seen as important. Naturally, there are also other ways to reward collaboration. We tackled this issue in our final question. The open question was formulated as follows: *"Would you like to collaborate more with other members of OI communities? How should groups be rewarded?"*

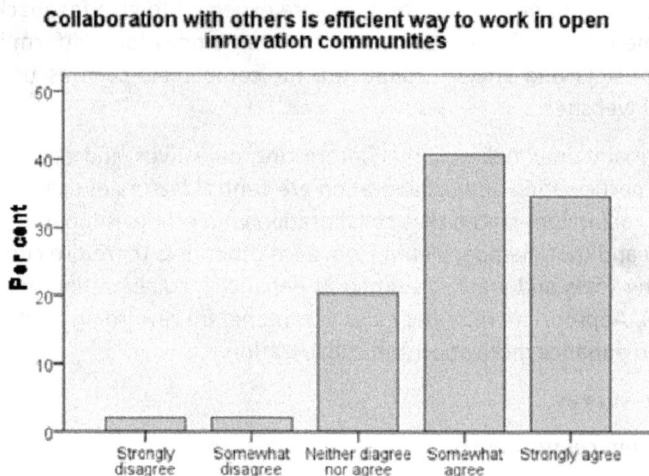

Figure 1: "What is important to you in an open innovation community?" N: 49, mean: 4.04, median: 4.00, std. dev.: 0.912

I would like to collaborate more with others in open innovation communities

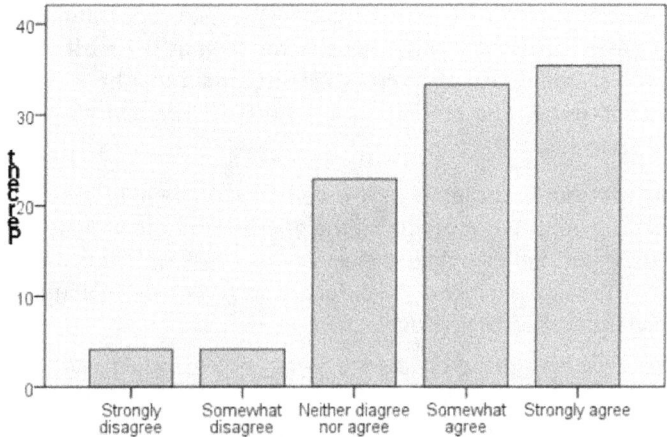

Figure 2: "What is important to you in an open innovation community?" N: 48, mean: 3.92, median: 4.00, std. dev.: 1.069

I would only collaborate with others if everyone in the group will be rewarded

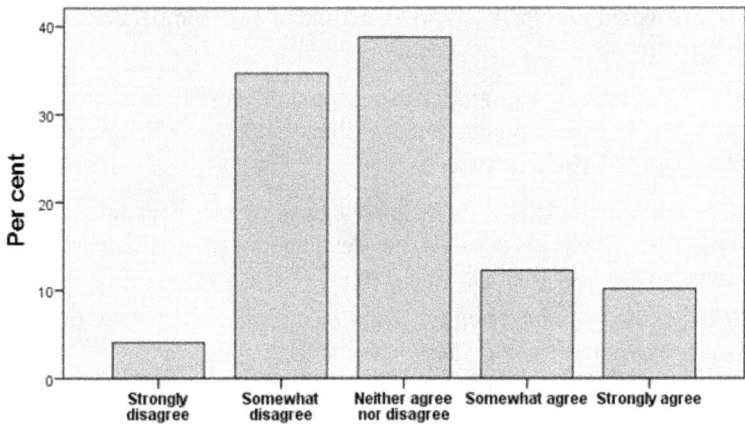

Figure 3: "What is your opinion on rewarding in an open innovation community?" N: 48, mean: 2.9 median: 3.00, std. dev.: 1.026

Once again, the answers indicate that collaboration was appreciated. Both online meetings and face-to-face meetings were suggested. Scheduled Internet sessions were seen as the preferable way to cooperate. Acquiring different perspectives from other people was seen as the most important benefit. For example, one of the respondents stated that he was an innovative person but not technically skilled. He therefore appreciated cooperation with technical people.

Members also saw a challenge in rewarding every member in the project equitably. The importance of finding the originator of the process was emphasised. One of the respondents suggested a point system that acknowledged the input of members at different stages of development, from problem identification to product launch.

> *"After launching the product, the same system would work as a royalty system that continued to reward the original contributors while allowing a traditional retail system to be imposed when marketing the product."*

A system in which the project leader distributes the rewards to his/her group was also suggested. The system would be transparent so that every member's scores were shown. This way, if the group leader is not equitable, he may not be voted leader again. Moreover, one idea was to choose a leader who would gain the biggest reward and all the members would then vote for the rewards.

Both kinds of rewards, monetary and recognition, were in fact seen as important. One of the respondents suggested that companies could invite members to visit their premises.

Furthermore, obtaining some kind of response to the ideas immediately and seeing how the ideas were further developed were emphasised in the members' answers, as the following statement indicates:

> *"If the quality of innovation produced by a group is high, that is its own reward, especially if the innovation is tangibly put into practice."*

One of the respondents indicated that monetary recognition should be assigned if the idea were to become commercially exploited. Moreover, well-timed rewarding was emphasized, as the following statement suggests:

"As the primary purpose of an innovation community is to develop new products that can then be released into the market, it is extremely important to reward the early contributors who develop the product to ensure the success and longevity of an innovation community."

3.4. Summary of findings in the context of the collective innovation process in OI communities

3.4.1. Motivation factors to collaborate

The results show that *interesting objectives* and *concepts* motivate users to collaborate in OI communities. One good example is hobbyist communities in which enthusiastic users can easily be motivated to participate and collaborate. An *open and constructive atmosphere* also motivates users to collaborate with others. The results show that users are willing to collaborate if they feel that they can *influence the product/service development.* Users also mentioned that they collaborated to *gain new viewpoints.* According to the results, users are motivated to collaborate because they consider it an *efficient way* to operate. On the other hand, from the hedonistic viewpoint, users find collaboration *fun.* Moreover, the *sense of cooperation and community* and *similarity* with other users also motivate users to collaborate. Finally, the right kind of *rewarding* that supports collaboration is an important motivator in the users' eyes.

3.4.2. Important elements of the rewarding strategy

First, the rewarding strategy should be *transparent and logical.* In other words, users should know why the rewards are given. Second, *democracy and equity* of the system are needed. Users also want the chance to influence the distribution of the rewards, for example, by voting. Every user should also feel that the system is fair. Third, *flexibility* of the strategy ensures that the nature of the rewards can vary in different situations. In, for example, the commercialisation phase, monetary rewards may be more significant. Intangible rewards, however, may support the aim of the fun aspect of the community. Fourth, *customisability* of the rewarding strategy ensures that users can influence, on some level, the nature of the rewards they receive and that the rewards will therefore be valuable to everyone. Finally, *active participation by the maintainers* is essential to the rewarding strategy. The results show that users want to receive feedback from the maintainers on their ideas. They also appreciate rewards such as visiting the maintainers' premises.

4. Discussion and conclusions

This explorative study is one of the first studies of collaboration in online OI communities presenting some preliminary results based on the interviews and the survey. In addition to serving academia, the study provides practical knowledge on how to reward and motivate groups of members on the web to companies and the growing number of OI intermediaries building or planning to build innovation communities. The study tackles two specific themes: users' motivations to collaborate and how rewarding can be used to motivate members to collaborate in OI communities.

Prior studies have shown that collaboration improves the quality of ideas by increasing the level of efficiency and creativity (e.g., Hargadon and Beckhy 2006; Thrift 2006). The study brought out all three kinds of motivation factors: intrinsic, extrinsic (Deci and Ryan 1985) and social (Motzek 2007).

Maintainers stated that the development of collaboration tools and methods is still more or less in its infancy. They admitted that users are currently working more as individuals than as groups. As one of the maintainers proposed, the next step would be for users to work genuinely as a group and not as individuals. This poses further challenges for tools and methods to be used for collaborative online innovation.

The second research question considered rewarding as a motivation factor. The importance of a well-designed reward system to the aim of receiving benefits can be logically justified based on prior literature. Within psychology, for example, the research by Deci (1975), Lerner and Tirole (2002) has presented results showing that expected monetary rewards tend to reduce intrinsic motivation whereas praise and other positive verbal feedback tend to increase it. In contrast, although intrinsic motivations seem to be important in the OI communities studied, the study emphasized both types of reward, monetary and non-monetary. The respondents to the study also gave concrete suggestions for the different kinds of rewarding models within groups. The results show that users are willing to collaborate for hedonistic and utilitarian reasons, and the collaboration possibilities in themselves can therefore be regarded as a reward. Furthermore, motivation factors to collaborate can and should be used when developing the rewarding strategy of the collaborative OI community. If, for example, users are motivated to collaborate because of the sense of community, a reward could be a visit to the company's premises, as suggested in this study. Furthermore, if fun is the motivator, rewards should somehow support this idea.

According to the results, a rewarding strategy should be transparent and logical. Rewarding should be based on the efforts and quality of the work rather than on the quantity of ideas or lotteries. Although users did not support the idea that everyone in the group be rewarded, they wanted to know the reasoning behind rewarding decisions. The system should also be flexible so that rewards vary in different situations and phases of the work. Equity and democracy of the rewarding system are important factors for OI community users. The customisability of the rewarding strategy ensures that users can influence, on some level, the nature of the rewards they receive, and that the rewards will therefore valuable to everyone.

This study clearly shows the untapped possibilities that lie in developing and enhancing collaboration in OI communities. Our interviews indicate that maintainers have recognized these possibilities at some point and are seeking solutions to support collaboration in different ways. This study brought out users' motivation factors to collaborate and important elements of the rewarding strategy that can be used together in the development of rewarding schemes in collaborative OI communities. There are opportunities for future studies to elaborate on the similarities and differences between the factors that determine generally appropriate rewarding strategies.

Acknowledgments

I would like to express my gratitude to Professor Saku Mäkinen at Tampere University of Technology for his valuable comments on this article. I am also grateful to my partner in the data collection and analysis, Heli Väätäjä, at Tampere University of Technology. I would further like to thank Ruben Robert from FellowForce, David Lionel from CrowdSpirit, and Asta Bäck and Pirjo Näkki from Owela/VTT for providing insightful information.

References

Ahonen, M., Antikainen, M., & Mäkipää, M. (2007) "Supporting collective creativity", European Academy of Management (EURAM) Conference proceedings, IDEAD, Paris.

Antikainen, M. and Väätäjä, H. (2010) "Rewarding in open innovation communities – How to motivate members?" *International Journal of Entrepreneurship and Innovation Management*, Vol 11, No. 4, pp 440-456.

Amabile, T. (1993) "Motivational synergy" *Human Resource Management Review*, No. 3, pp 185-201.

Amabile, T. (1996) *Creativity in Context*, Boulder.

Bagozzi, R. and Dholakia, U. (2002) "Intentional social action in consumer behaviour", *Journal of Interactive Marketing*, Vol 16, No. 2, pp 2-21.

Bandura, A. (2005) *Self-efficacy in changing societies*, Cambridge University Press, New York.

Chesbrough, H.W. (2003) Open Innovation: The New Imperative for Creating and Profiting from Technology, Harvard Business School Press, Boston, MA.

Chesbrough, H. (2006) "Open Innovation: A Paradigm for Understanding Industrial Innovation", In Chesbrough, H., Vanhaverbeke, W., & West, J. (Eds.) *Open Innovation: Researching a NewParadigm* 134, Oxford University Press, Oxford, UK.

Chesbrough, H. (2006a) *Open Business Models: How to Thrive in the New Innovation Landscape*, Harvard Business School Press, Boston, MA.

Constant, D., Kiesler, S and Sproull, L. (1994) "What's mine is ours, or is it? A study of attitudes about information sharing", *Information Systems Research*, Vol 5, No. 4, pp 400–421.

Deci, E. (1975) *Intrinsic Motivation*, New York.

Deci, E. and Ryan, R. (1985) Intrinsic Motivation and Self-Determination in Human Behaviour, New York and London.

Gruen, T., Osmonbekov, T. and Czaplewski, A. (2005) "How e-communities extend the concept of exchange in marketing: An application of the motivation, opportunity, ability (MOA) theory", *Marketing Theory*, Vol 5, No. 1, pp 33-49.

Hagel, J. and Armstrong, A. (1997) *Net gain: Expanding markets through virtual communities*, McKingsey & Company, Boston.

Hargadon, A. and Bechky, B. (2006) "When collections of Creatives Become Creative Collective – a Field Study of Problem Solving at Work", *Organisation Science*, Vol 17, No. 4, pp 484-500.

Harper, F.M., Raban, D., Rafaeli, S. and Konstan, J.A. (2008) "Predictors of answer quality in online Q&A sites", *Proceedings of the twenty-sixth annual SIGCHI conference on Human factors in computing systems*, ACM, New York, USA, pp 865-874.

Hars, A and Ou, S. (2002) "Working for free? Motivations for participating in open source Projects", *International Journal of Electronic Commerce*, Vol 6, No. 3, pp 25–39.

Jeppesen, L., Frederiksen, L. (2006) "Why do users contribute to firm-hosted user communities? The Case of Computer-Controlled Music Instruments", *Organisation Science*, Vol 17, No. 1, January–February 2006, pp 45–63.

Kittur, A., Chi, E. and Bongwon, S. (2008) "Crowdsourcing user studies with mechanical turk", *Proceedings of the twenty-sixth annual SIGCHI conference on Human factors in computing systems*, ACM, New York, USA, pp 453-456.

Kollock, P. (1999) "The economies of online cooperation: Gifts and public goods in cyberspace", in M. Smith and P. Kollock (eds), *Communities in cyberspace*, Routledge, London.

Lakhani, K. R. and Wolf, R. (2005) "Why hackers do what they do: understanding motivation and effort in free/open source software projects", in (Eds.) J. Feller, S. Hissan and K.R. Lakhani, *Perspectives on free and open source software*, MIT Press.

Lepper, M., Greene, D. and Nisbett R. (1973) "Undermining children's intrinsic interest with extrinsic rewards: A test of the 'overjustification' hypothesis", *Journal of Personality and Social Psychology*, 28, pp 129-137.

Lerner, J., and Tirole, J. (2002) "Some simple economics of open source", *Journal of Industrial Economics*, Vol 50, No. 2, pp 197-234.

Lindenberg, S. (2001) *"Intrinsic motivation in a new light"*, *Kyklos*, Vol 54, No. 2-3, pp 317–342.

Motzek, R. (2007) Motivation in Open Innovation. An exploratory study on User Innovators, VDM Verlag, Saarbrücken.

Nov, O. (2007) "What motivates Wikipedians?", *Communications of the ACM*, Vol 50, No. 11, pp 60-64.

Osterloh, M., Rota, S, Kuster, B. (2004) "Open Source Software Production: Climbing on the Shoulders of Giants", Zurich 2004, [online], Retrieved April 14, 2009 from http://opensource.mit.edu/papers/osterlohrotakuster.pdf

Raymond, E. S. (2001) *The cathedral and the bazaar*, O'Reilly, Sebastopol, CA.

Rheingold, H. (1993) The virtual community: Homesteading on the electronic frontier, Addison-Wesley, New York.

Ridings, C. and Gefen, D. (2004) "Virtual community attraction: Why people hang out online", *Journal of Computer-Mediated-Communication*, Vol 10, No. 1, Article 4.

Reeve, J. (2005) *Understanding motivation and emotion,* John Wiley & Sons, Inc, USA, 556 p.

Ryan, R., and Deci, E. (2000) "Self-determination theory and the facilitation of intrinsic motivation, social development, and well-being", *American Psychologist*, Vol. 5, No. 1, pp 68-78.

Stewart, K. and Gosain, S. (2006) "The impact of ideology on effectiveness in open source software development teams" *MIS Quarterly,* Vol 30, No. 2. pp 291–314.

Thrift, N. (2006) Re-inventing invention: new tendencies in capitalist commodification. Economy and Society, Vol 35, No. 2, pp 279-306.

Torvalds, L., & Diamond, D. (2001) *Just for fun: The story of an accidental revolutionary*, Harper Business, New York.

von Hippel, E. (2005) *Democratizing innovation,* The MIT Press.

von Hippel, E. and von Krogh, G. (2003) "Open source software and the 'private-collective' innovation model: Issues for organization science", *Organization Science*, Vol 14, No. 2, pp 208–223.

Wasko, M. and Faraj, S. (2000) "It is what one does: why people participate and help others in electronic communities of practice", *Journal of Strategic Information Systems*, Vol 9, No. 2-3, pp 155-173.

Wiertz, C. and de Ruyter, K. (2007) "Beyond the call of duty: Why customers contribute to firm-hosted commercial online communities", *Organization Studies*, Vol 28, No. 3, pp 347-376.

Zeityln, D. (2003) "Gift economies in the development of open source software: Anthropological reflections", *Research Policy*, Vol 32, pp 1287-1291.

www.ingramcontent.com/pod-product-compliance
Lightning Source LLC
Chambersburg PA
CBHW061212220326
41599CB00025B/4619